Construction

Volume 3

Second Edition

R. CHUDLEY MCIOB
Chartered Builder

Illustrated by the author

MARINA VILLAFANA

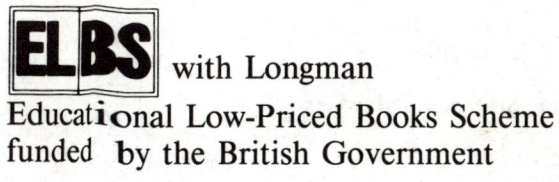

ELBS with Longman
Educational Low-Priced Books Scheme
funded by the British Government

VICTOR MANHIN LTD.
49 HIGH STREET
SAN FERNANDO
TRINIDAD W.I.

Addison Wesley Longman Limited
Edinburgh Gate, Harlow,
Essex CM20 2JE, England

Associated companies throughout the world

© Longman Group Ltd 1976, 1987

The Educational Low-Priced Books Scheme is funded by the Overseas Development Administration as part of the British Government overseas aid programme. It makes available low-priced, unabridged editions of British publishers' textbooks to students in developing countries.

All rights reserved; no part of this publication may be reproduced, stored in a retrieval system, or transmitted in any form or by any means, electronic, mechanical, photocopying, recording, or otherwise, without the prior written permission of the Publishers.

First published 1976
Eighth impression 1985
Second edition 1987
Reprinted, 1991, 1992, 1994, 1997

ELBS edition first published 1978
Reprinted 1980, 1983, 1985
ELBS edition of second edition 1987
Reprinted 1989, 1991, 1992, 1994, 1995, 1996, 1997

ISBN 0 582 32282 0

Set in IBM Journal 10 on 12 point

Produced by Longman Singapore Publishers (Pte) Ltd
Printed in Singapore

Contents

Introduction		v
Part I Site works		1
1	Deep trench excavations	1
2	Tunnelling	14
3	Demolition	24
Part II Foundations		30
4	Underpinning	30
5	Piled foundations	41
Part III Frameworks		65
6	Portal frame theory	65
7	Concrete portal frames	69
8	Steel portal frames	76
9	Timber portal frames	81
Part IV Fire		87
10	The problem of fire	87
11	Structural fire protection	90
12	Means of escape in case of fire	117
Part V Claddings to framed structures		141
13	Cladding panels	141
14	Infill panels	147
15	Jointing	152
16	Mastics and sealants	160

Part VI	Factory buildings	163
17	Roofs	163
18	Walls	177
19	Wind pressures	183
Part VII	Formwork	188
20	Wall formwork	188
21	Patent formwork	199
22	Concrete surface finishes	208
Part VIII	Stairs	215
23	Concrete stairs	215
24	Metal stairs	231
	Bibliography	239
	Index	241

Acknowledgements

We are grateful to the following for permission to reproduce copyright material:

British Standards Institution for reference to British Standards, Codes of Practice; Building Research Establishment for extracts from Building Research Establishment Digests; Her Majesty's Stationery Office for extracts from Acts, Regulations and Statutory Instruments.

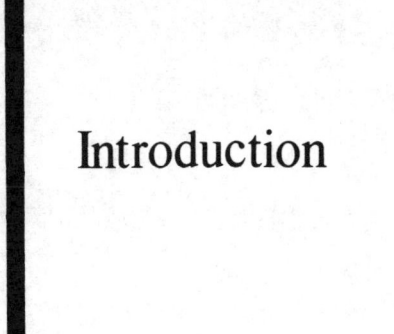

Introduction

The third year of a course of construction technology is sometimes called the first year of advanced technology. It is very difficult to make a clear and concise definition between the terms 'elementary' and 'advanced' technology. A better approach is to consider each year's work as being a development of previous years' studies, progressing from the relatively simple domestic structure to the more complex forms of construction.

As structures become larger and consequently more complicated the services involved need special consideration. It is general in courses of third-year standard to study services as a separate topic under the guidance of a specialist; for this reason, unlike the previous volumes, services have been omitted from this text.

The format adopted follows that of Volumes 1 and 2, being in concise note form amply illustrated to elaborate upon the text content. The reader should appreciate that the illustrations used are to emphasise a point of theory and must not be accepted as the only solution. A study of working drawings and details from building appraisals given in the various journals will add to the student's background knowledge and comprehension of constructional technology.

1
Deep trench excavations

Part I
Site works

Any form of excavation on a building site is a potential hazard, and although statistics show that of the 46 000 or so reportable accidents occurring each year on building sites excavations do not constitute the major hazard, but they can often prove to be serious; indeed, approximately 1 in 10 accidents occurring in excavations are fatal.

THE CONSTRUCTION (GENERAL PROVISIONS) REGULATIONS 1961

This is a statutory instrument made under the powers of the Factories Acts of 1937 and 1948. They were introduced to provide a minimum degree of safety to operatives when working in or near excavations, shafts, tunnels, demolitions, work involving the use of explosives and work on or adjacent to water. These regulations apply to building operations and works of engineering construction since the risks encountered in the two industries are similar and therefore a common code of safety is desirable.

In general the regulations set out the requirements for the supply and use of adequate timbering to excavations, the appointment of safety supervisors for firms employing a total of more than 20 persons engaged on constructional works and the frequency of inspections to ascertain the safe condition of the working areas.

Every contractor is responsible for the safety of his own employees and every person employed must co-operate in observing the various requirements of the regulations. If an employee discovers any defect or unsafe condition in a working area it is his duty to report the facts to his employer, foreman or to a person appointed by the employer as safety supervisor. The requirements embodied in the Health and Safety at Work, etc., Act 1974 must also be observed.

DEEP TRENCH EXCAVATIONS

Deep trenches may be considered as those over 3.000 m deep and are usually required for the installation of services since deep foundations are not very often encountered due to the more economic alternatives such as piling and raft techniques. Trench excavations should not be opened too far in advance of the proposed work and any backfilling should be undertaken as soon as practicable after the completion of the work. These two precautions will lessen the risk of falls, flooding and damage to completed work as well as releasing the timbering for re-use at the earliest possible date. Great care must be taken in areas where underground services are present; these should be uncovered with care, protected and supported as necessary. The presence of services, in an excavation area, may restrict the use of mechanical plant to the point where its use becomes uneconomic. Hand trimming should be used for bottoming the trench, side trimming, end trimming and for forming the gradient just prior to the pipe, cable or drain laying.

In general all deep excavations should be close boarded or sheeted as a precautionary measure, the main exception being hard and stable rock subsoils. Any excavation in a rock strata should be carefully examined to ascertain its stability. Fissures or splits in the rock layers which slope towards the cut face may lead to crumbling or rock falls particularly when exposed to the atmosphere for long periods. In this situation it would be prudent to timber the faces of the excavation according to the extent and disposition of the fissures.

In firm subsoils it might be possible to complete the excavation dig before placing the timbering in position. The general method of support for the excavation sides follows that used in shallow and medium depth trench excavations studied in the first 2 years of a construction technology course, except that larger sections are used to resist the increased pressures encountered as the depth of excavation increases — see Fig. I.1

If the subsoil is weak, waterlogged or running sand it will be necessary to drive timber runners, trench sheeting or interlocking sheet piles of steel, timber or precast concrete ahead of the excavation dig. This can be accomplished by driving to a depth exceeding the final excavation depth or by the drive and dig system ensuring that timbering is always in advance of the excavating operation. Long runners or sheet piles will require a driving frame to hold and guide the members whilst being driven. To avoid the use of large piling frames and heavy driving equipment the

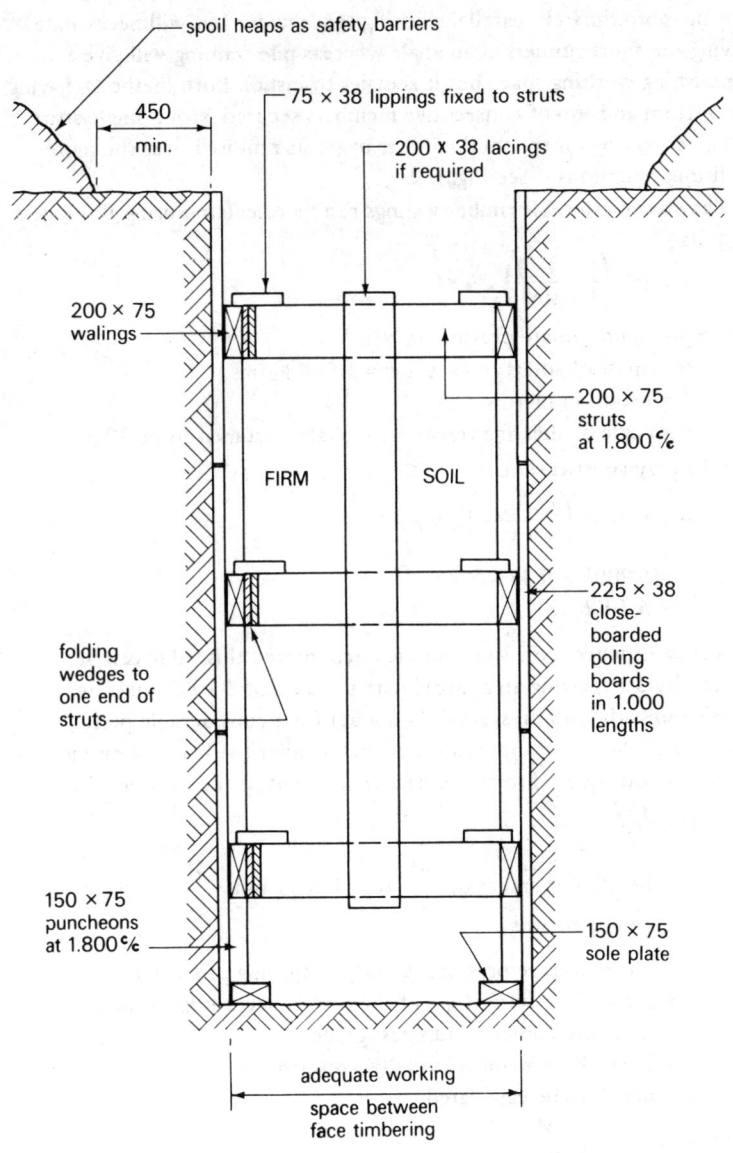

Fig. I.1 Traditional deep trench timbering

tucking and pile framing techniques may be used. Tucking framing will give an approximately parallel working space width but will necessitate driving the short runners at an angle whereas pile framing will give a diminishing working space but it is easier to install. Both methods, having the bottom and top of consecutive members secured with a single strut, will give a saving on the total number of struts required over the more traditional methods — see Fig. I. 2.

The sizes of suitable timber walings can be calculated using Rankine's formula:

$$p = wh \left(\frac{1 - \sin\theta}{1 + \sin\theta}\right) \times 9.81$$

where p = approximate pressure in N/m^2
w = mass of soil (typical value = 1 980 kg/m^3)
h = height in metres
e = angle of shearing resistance (usually assumed to be 30°)

therefore by substitution:

$$p = 1\,980h \left(\frac{1 - 0.5}{1 + 0.5}\right) \times 9.81$$

$$= 1\,980h \times 0.33 \times 9.81$$

$$\simeq 6\,400h$$

in most cases where cohesive soils are encountered this value can be reduced by 50%, giving an approximate pressure of 3 200h, since in cohesive soils the full pressure does not act for a considerable period.

To select the actual dimensions of the member the bending moment is calculated and equated to the resistance moment of the member thus:

$$M = \frac{fbd^2}{6}$$

where M = bending moment of $\frac{wl^2}{10}$ for runners and $\frac{wl^2}{8}$ for short poling boards.

(w = average pressure 'p' and l = spacing of struts)
f = permissible fibre stress of the timber according to species, moisture content and stress grade
b = breadth of member (usually assumed)
d = depth to be calculated

therefore $d = \sqrt{\frac{6M}{fb}}$

It must be remembered that timbering is a general term used to cover all forms of temporary support to the sides of an excavation to:
1. Prevent collapse of the excavation sides and so endanger the operatives working in the immediate vicinity.

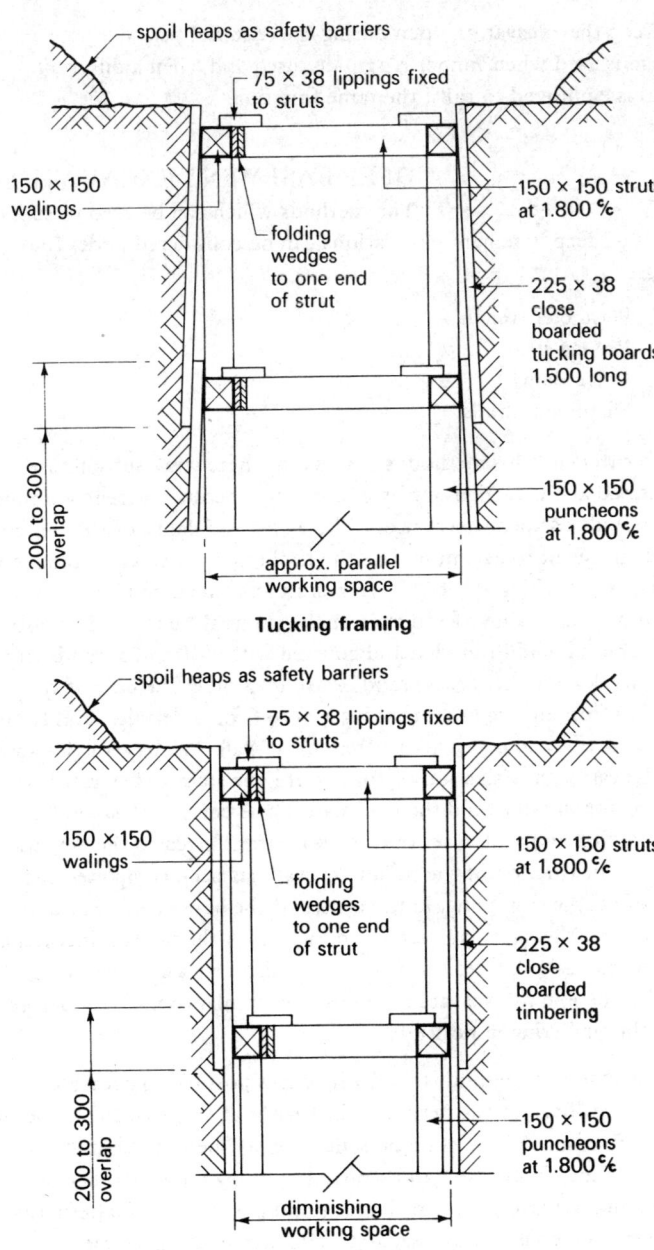

Fig. I.2 Timbering – tucking and pile framing

2. Keep the excavation open during the required period.

The term is used when timber is actually used and when a different material is employed to fulfil the same function.

DEEP BASEMENT EXCAVATIONS

The methods which can be used to support the sides of deep basement excavations can be considered under four headings:

1. Perimeter trench.
2. Raking struts.
3. Cofferdams.
4. Diaphragm walls.

Perimeter trench: this method is employed where weak subsoils are encountered and is carried out by excavating a perimeter trench around the proposed basement excavation. The width and depth of the trench must be sufficient to accommodate the timbering, basement retaining wall and adequate working space. The trench can be timbered by any of the methods described above for deep trenches in weak subsoils. The bottom of the trench should be graded and covered with a 50 to 75 mm blinding layer of weak concrete, coarse sand or ash to protect the base of the excavation from drying and shrinking and to form a definite level from which to set out and construct the basement wall. The base of the wall should be cast first with a kicker formed for the stem and starter bars left projecting for the stem and the base slab. The stem or wall should be cast in suitable lifts and as it cures the struts are transferred to the new wall. When the construction of the perimeter wall has been completed the mound of soil, or dumpling, in the centre of the basement area can be excavated and the base slab cast — see Fig. I.3. Although this method is intended primarily for weak subsoils it can also be used in firm soils where it may be possible to excavate the perimeter trench completely before placing the timbering in position.

Raking struts: this method is used where it is possible to excavate the basement area back to the perimeter line without the need for timbering, therefore firm subsoils must be present. The perimeter is trimmed and the timbering placed in position and strutted by using raking struts converging on a common grillage platform similar to raking shoring, or alternatively each raker is taken down to a separate sole plate and the whole arrangement adequately braced. An alternative method is to excavate back to the perimeter line on the subsoil's natural angle of repose and then cast the base slab to protect the excavation bottom from undue drying out and

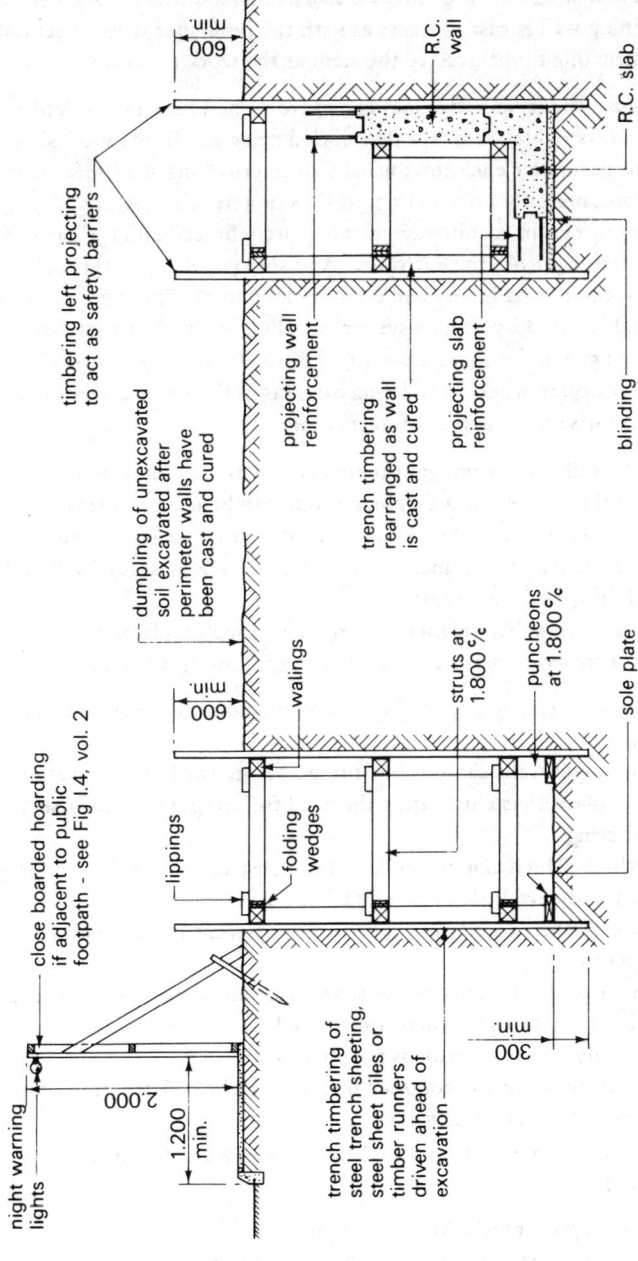

Fig. I.3 Basement timbering – perimeter trench method

subsequent shrinkage. The perimeter trimming can now be carried out, timbered and strutted using the base slab as the abutment — see Fig. I.4. The retaining wall is cast in stages as with the perimeter trench method and the strutting transferred to the stem as the work proceeds.

Cofferdams: the term cofferdam comes from the French word 'coffre' meaning a box, which is an apt description since a cofferdam consists of a watertight perimeter enclosure, usually of interlocking steel sheet piles, used in conjunction with waterlogged sites or actually in water. The enclosing cofferdam is constructed, adequately braced and the earth is excavated from within the enclosure. Any seepage of water through the cofferdam can normally be handled by small pumps. The sheet piles can be braced and strutted by using a system of raking struts, horizontal struts or tie rods and ground anchors — see Fig. I.5. Cofferdams are usually studied to a greater degree when considering caissons in the second year of an advanced course in construction technology.

Diaphragm walls: a diaphragm can be defined as a dividing membrane and in the context of building a diaphragm wall can be used as a retaining wall to form the perimeter wall of a basement structure, to act as a cut-off wall for river or similar embankments and to retain large masses of soil such as a side wall of a road underpass.

In situ concrete diaphragm walls are being used to a large extent in modern construction work and can give the following advantages:

1. Final wall can be designed and constructed as the required structural wall.
2. Diaphragm walls can be constructed before the bulk excavation takes place thus eliminating the need for temporary works such as timbering.
3. Methods which can be employed to construct the wall are relatively quiet and have little or no vibration.
4. Work can be carried out immediately adjacent to an existing structure.
5. They may be designed to resist vertical and/or horizontal forces.
6. Walls are watertight when constructed.
7. Virtually any plan shape is possible.
8. Overall they are an economic method for the construction of basement or retaining walls.

There are two methods by which a cast *in situ* diaphragm wall may be constructed:

1. Touching or interlocking bored piles.
2. Excavation of a trench using the bentonite slurry method.

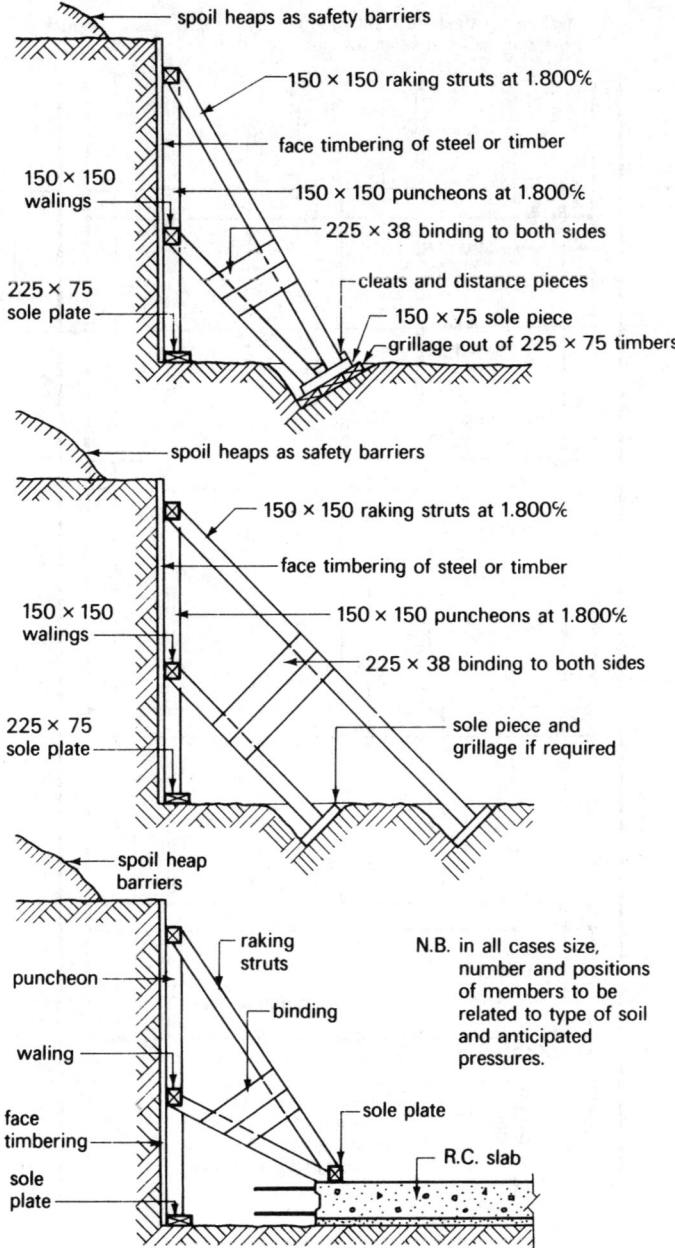

Fig. I.4 Basement timbering - raking struts

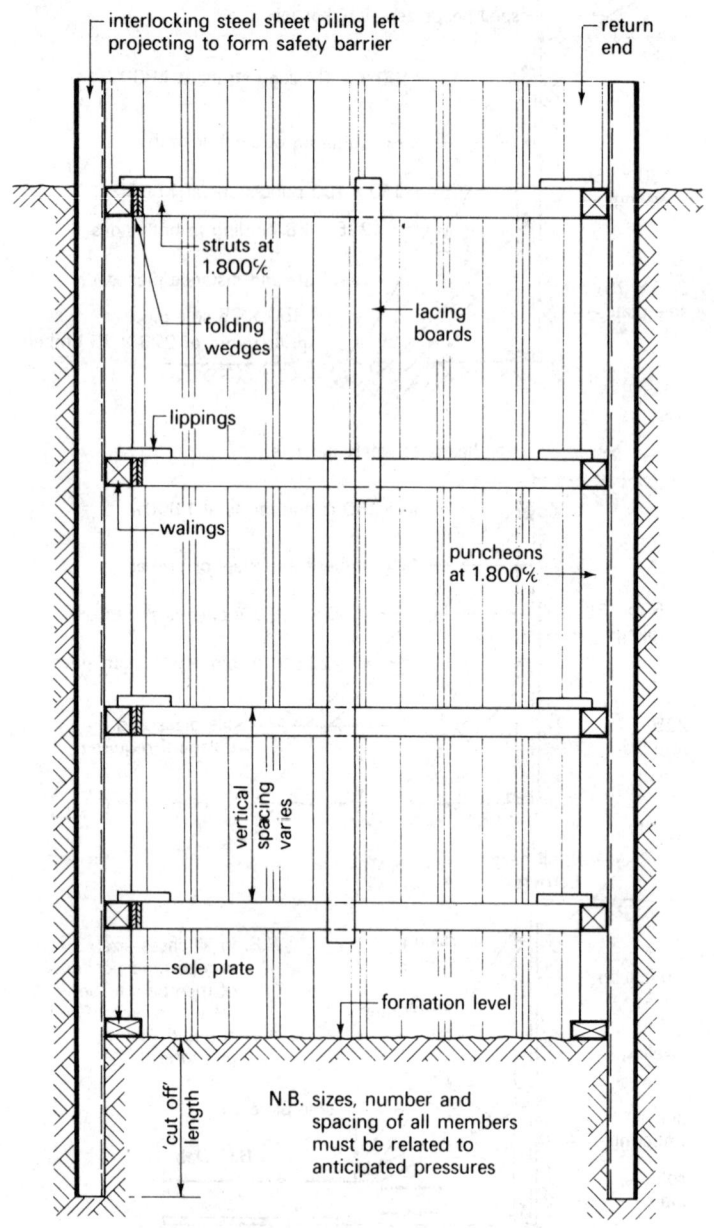

Fig. I.5 Cofferdams − basic principles

The formation of bored piles is fully described later in the chapter on piling. The piles can be bored so that their interfaces are just touching; or by boring for piles in alternate positions and using a special auger the intermediate pile positions can be bored so that the interfaces interlock. The main advantage of using this method is that the wall can be formed within a restricted headroom of not less than 2.000 m. The disadvantages are the general need for a reinforced tie beam over the heads of the piles and the necessity for a facing to take out the irregularities of the surface. This facing usually takes the form of a cement rendering lightly reinforced with a steel welded fabric mesh.

The general method used to construct diaphragm walls is the bentonite slurry system. Bentonite is manufactured from a montmorillonite clay which is commonly called fullers' earth because of its use by fullers in the textile industry to absorb grease from newly woven cloth. When mixed with the correct amount of water bentonite produces thixotropic properties giving a liquid behaviour when agitated and a gel structure when undisturbed.

The basic procedure is to replace the excavated spoil with the bentonite slurry as the work proceeds. The slurry forms a soft gel or 'filter cake' at the interface of the excavation sides with slight penetration into the subsoil. Hydrostatic pressure caused by the bentonite slurry thrusting on the 'filter cake' cushion is sufficient to hold back the subsoil and any ground water which may be present. This alleviates the need for timbering and/or pumping and can be successfully employed up to 36.000 m deep.

Diaphragm walls constructed by this method are executed in alternate panels from 4.500 m to 7.000 m long with widths ranging from 500 to 900 mm using a specially designed hydraulic grab attached to a standard crane or by using a continuous cutting and recirculating machine. Before the general excavation commences a guide trench, about 1.000 m deep is excavated and lined with lightly reinforced walls. These walls act as a guide line for the excavating machinery, provide a reservoir for the slurry, enables pavings and underground services to be broken out ahead of the excavation.

To form an interlocking and watertight joint at each end of the panel, circular stop end pipes are placed in the bentonite-filled excavation before the concrete is placed. The continuous operation of concreting the panel is carried out using a tremie pipe and a concrete mix designed to have good flow properties without the tendency to segregate. This will require a concrete with a high slump of about 200 mm but with high strength properties ranging from 20 to 40 N/mm^2. Generally the rate of pour is in the region of 15 to 20 m^3 per hour and as the concrete is introduced into the excavated panel it will displace the bentonite slurry, which is less dense

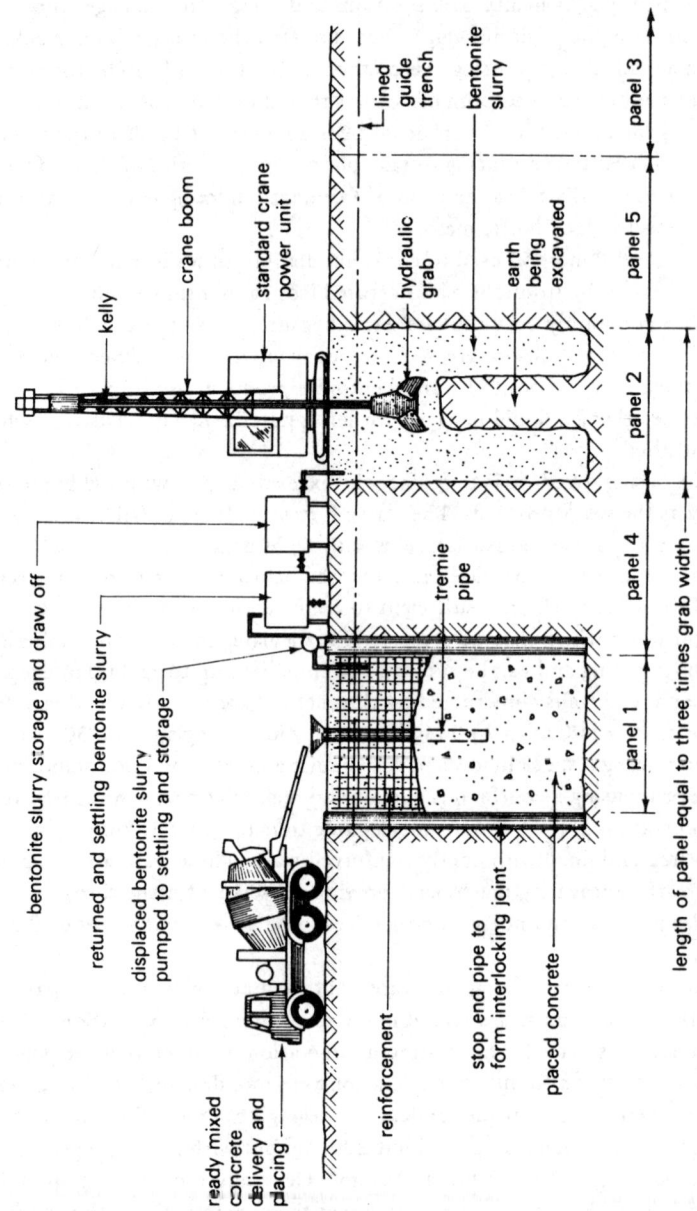

Fig. 1.6 Diaphragm wall construction – bentonite slurry method

than the concrete, which can be stored for re-use or transferred to the next panel being excavated. The ideal situation is to have the two operations acting simultaneously and in complete unison — see Fig. I.6. Before the concrete is placed reinforcement cages of high yield or mild steel bars are fabricated on site in one or two lengths. Single length cages of up to 20.000 m are possible; where cages in excess of this are required they are usually spot welded together when the first cage is projecting about 1.000 m above the slurry level. The usual recommended minimum cover is 100 mm which is maintained by having spacing blocks or rings attached to the outer bars of the cage. Upon completion of the concreting the bentonite slurry must be removed from the site either by tanker or by diluting so that it can be discharged into the surface water sewers by agreement with the local authority.

2
Tunnelling

A tunnel may be defined as an artificial underground passage and has been used by man as a means of communication or for transportation for several thousand years. Prehistoric man is known to have connected his natural cave habitats by tunnels hewn in the rock. A Babylonian king *circa* 2180-2160 B.C. connected his royal palace to the Temple of Jupiter on the opposite bank of the Euphrates by a brick arched tunnel under the river. Other examples of early tunnels are those hewn in the rock in the tomb of Mineptah at Thebes in Egypt and the early Greek tunnel, constructed about 687 B.C., used for conveying water on the Island of Samos.

The majority of these early tunnels were constructed in rock subsoils and therefore required no permanent or temporary support. Today, tunnelling in almost any subsoil is possible. Permanent tunnels for underground railways and roads can be lined with metal and/or concrete but such undertakings are the province of the civil engineer. The general building contractor would normally only be involved with temporary tunnelling for the purposes of gaining access to existing services or installing new services, constructing small permanent tunnels for pedestrian subways under road or railway embankments and forming permanent tunnels for services.

When the depth of a projected excavation is about 6.000 m the alternative of working in a heading or tunnel should be considered taking into account the following factors:

1. *Nature of subsoils* — the amount of timbering that will be required in the tunnel as opposed to that required in deep trench excavations.

2. *Depth of excavation* — over 9.000 m deep it is usually cheaper to tunnel or use one of the alternative methods such as thrust boring. The cover of ground over a tunnel to avoid disturbance of underground services, roads, pavings and tree roots is generally recommended to be 3.000 m minimum.
3. *Existing services* — in urban areas buried services can be a problem with open deep-trench excavations; this can generally be avoided by tunnelling techniques.
4. *Carriageways* — it may be deemed necessary to tunnel under busy roads to avoid disturbance of the flow of traffic.
5. *Means of access* — the proposed tunnel may be entered by means of an open trench if the tunnel excavation is into an embankment or access may be gained by way of a shaft.
6. *Construction Regulations* — Part IV of the Construction (General Provisions) Regulations 1961 sets out the minimum requirements for the protection of operatives working in excavations, shafts and tunnels covering such aspects as temporary timbering, supervision of works and means of egress in case of flooding. Part VII deals with the ventilation of excavations and Regulation 47 in Part XI covers the provision of adequate lighting.

SHAFTS

These are by definition vertical passages but in the context of building operations they can also be used to form the excavation for a deep isolated base foundation. In common with all excavations, the extent and nature of the temporary support or timbering required will depend upon:
1. Subsoil conditions encountered.
2. Anticipated ground and hydrostatic pressures.
3. Materials used to provide temporary support.
4. Plan size and depth of excavation.

In loose subsoils a system of sheet piling could be used by driving the piles ahead of the excavation operation. This form of temporary support is called a cofferdam and is fully covered in the second year of advanced study — see Fig. I.5.

Alternatives to the cofferdam techniques for shaft timbering are tucking framing and pile framing. These methods consist of driving short timber runners, 1.000 to 2.000 m long, ahead of the excavation operation and then excavating and strutting within the perimeter of the runners. The process is repeated until the required depth has been reached. It is essential with all drive and dig methods that at all times the depth to which the runner has been driven is in excess of the excavation depth. The installation of tucking framing and pile framing in shafts is as described for deep trench timbering — see Fig. I.2. Both tucking framing and pile framing have the advantages over sheet piling of not requiring large guide trestles and heavy driving equipment.

In firm subsoils the shaft excavation would be carried out in stages of 1.000 to 2.000 m deep according to the ability of the subsoil to remain stable for short periods. Each excavated stage is timbered before proceeding with the next digging operation. The sides of the excavated shaft can be supported by a system of adequately braced and strutted poling boards — see Figs I.7 and I.8. Sometimes a series of cross beams are used at the head of shaft timbering to reduce the risk of the whole arrangement sliding down the shaft as excavation work is proceeding at the lower levels.

Shafts are usually excavated square in plan with side dimensions of 1.200 to 3.000 m depending on:
1. Total depth required.
2. Method of timbering.
3. Sizes of support lining to save unnecessary cutting to width.
4. Number of operatives using or working within the shaft.
5. Size of skip or container to be used for removing spoil.
6. Type of machinery used for bulk excavation.

If the shaft is for the construction of an isolated base then an access or ladder bay should be constructed. This bay would be immediately adjacent to the shaft and of similar dimensions, making in plan a rectangular shaft. The most vulnerable point in any shaft timbering is the corners, where high pressures are encountered, and these positions should be specially strengthened by using corner posts or runners of larger cross section — see Fig. I.7.

TUNNELS

The operational sequence of excavating and timbering a tunnel or heading by traditional methods can be enumerated thus:
1. In firm soils excavate first 1.000 m long stage or bay; if weak subsoil is encountered it may be necessary to drive head boards and lining boards as the first operation.
2. Head boards 1.000 m long are placed against the upper surface.
3. Sole plate and stretcher are positioned; these are partly bedded into the ground to prevent lateral movement and are levelled through from stretcher to stretcher.
4. Cut and position head tree.

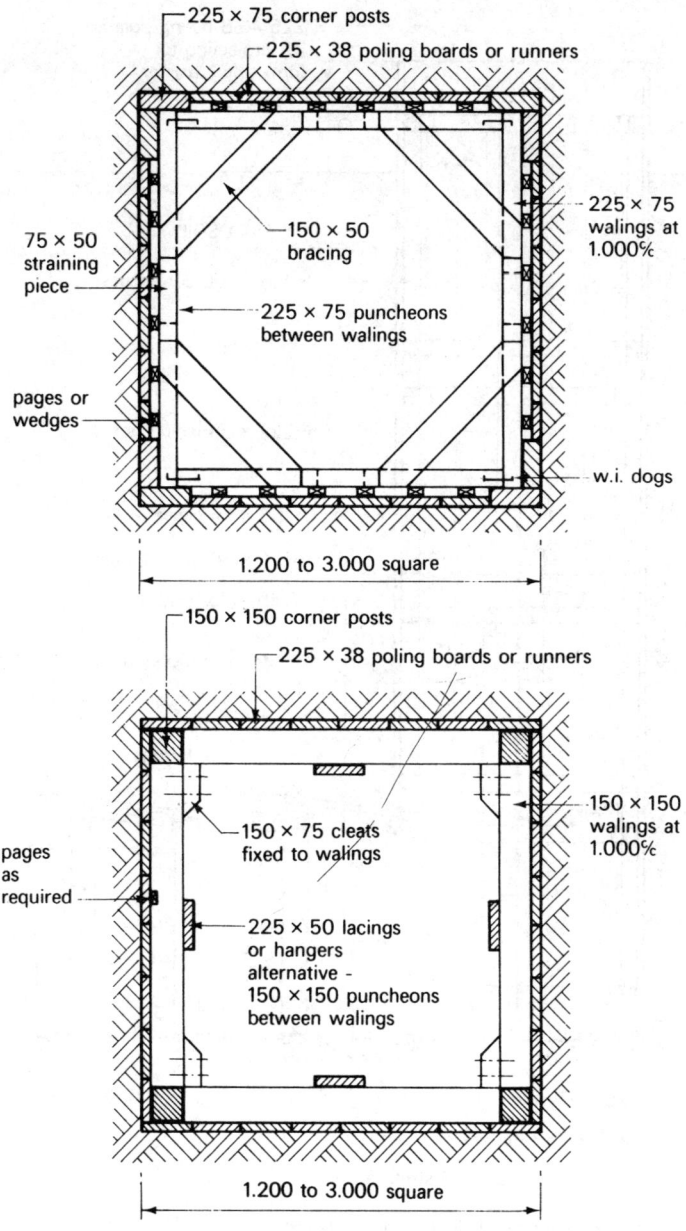

Fig. I.7 Shaft timbering – typical plans

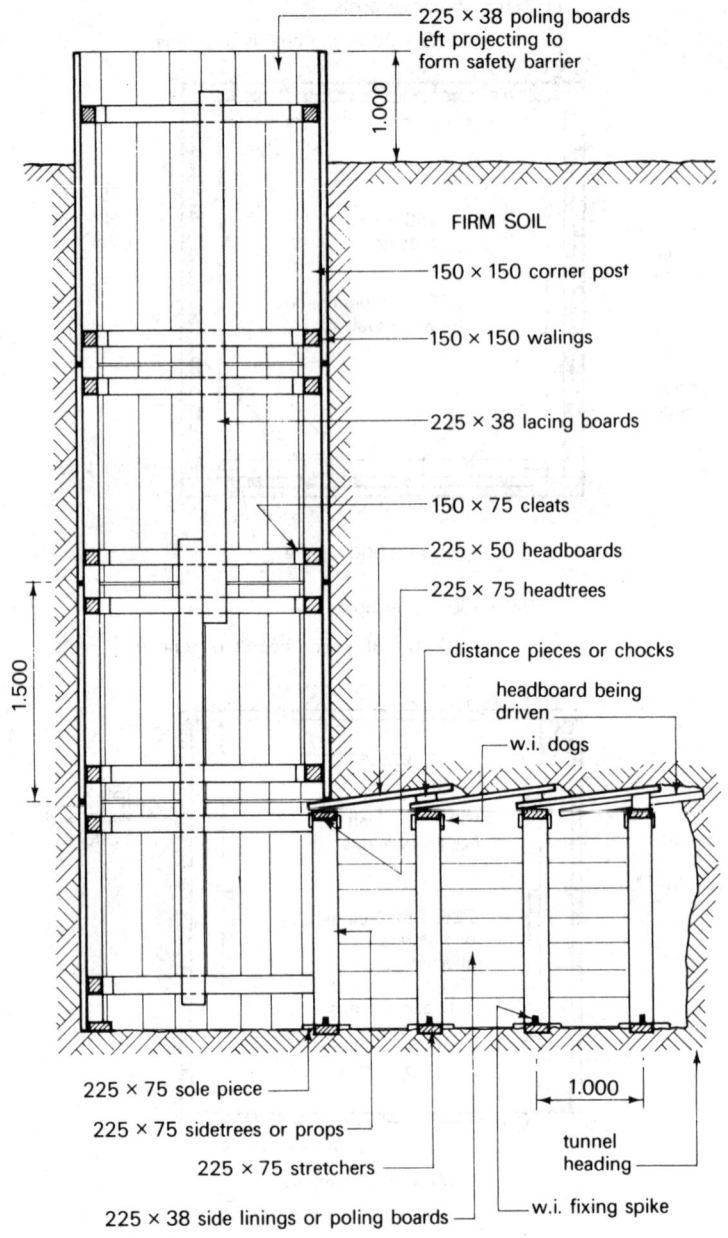

Fig. I.8 Typical shaft and tunnel timbering

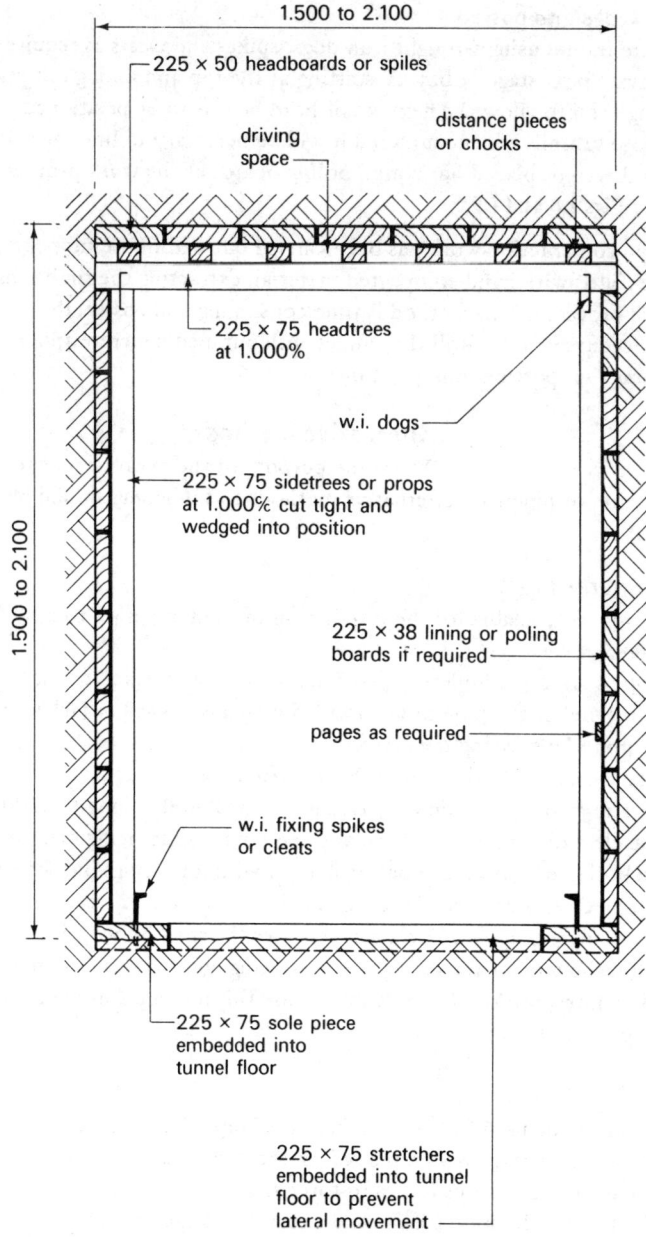

Fig. I.9 Typical section through tunnel timbering

5. Cut side trees or struts to fit tightly between sole plate and head tree and wedge into position.
6. Secure frames using wrought iron dogs, spikes and cleats as required.
7. Excavate next stage or bay by starting at the top and taking out just enough soil to allow the next set of head boards to be positioned.
8. If loose subsoils are encountered it will be necessary to line the sides with driven or placed horizontal poling boards as the work proceeds — see Figs I.8 and I.9.

After the construction work has been carried out within the tunnel it can be backfilled with hand-compacted material, extracting the timbering as the work proceeds. This method is time consuming and costly; the general procedure is to backfill the tunnel with pumped concrete and leave the temporary support work in position.

Alternative methods

Where the purpose of the excavation is for the installation of pipework alternative methods to tunnelling should be considered.

Small diameter pipes
Two methods are available for the installation of small pipes up to 200 mm diameter.
1. *Thrust boring* — a bullet-shaped solid metal head is fixed to the leading end of the pipe to be installed which is pushed or jacked into the ground displacing the earth.
2. *Auger boring* — carried out with a horizontal auger boring tool operating from a working pit having at least 2.400 m long × 1.500 m wide clear dimensions between any temporary supporting members. The boring operation can be carried out without casings but where the objective is the installation of services, concrete or steel casings are usually employed. The auger removes the spoil by working within the bore of the casing which is being continuously rammed or jacked into position. It is possible to use this method for diameters of up to 1.000 m.

Pipe jacking
This method can be used for the installation of pipes from 150 to 3 600 mm diameter but it is mainly employed on the larger diameters of over 1.000 m. Basically the procedure is to force the pipes into the subsoil by means of a series of hydraulic jacks and excavate, as the driving proceeds, from within the pipes by hand or machine according to site conditions. The leading pipe is usually fitted with a steel shield or hood to aid the

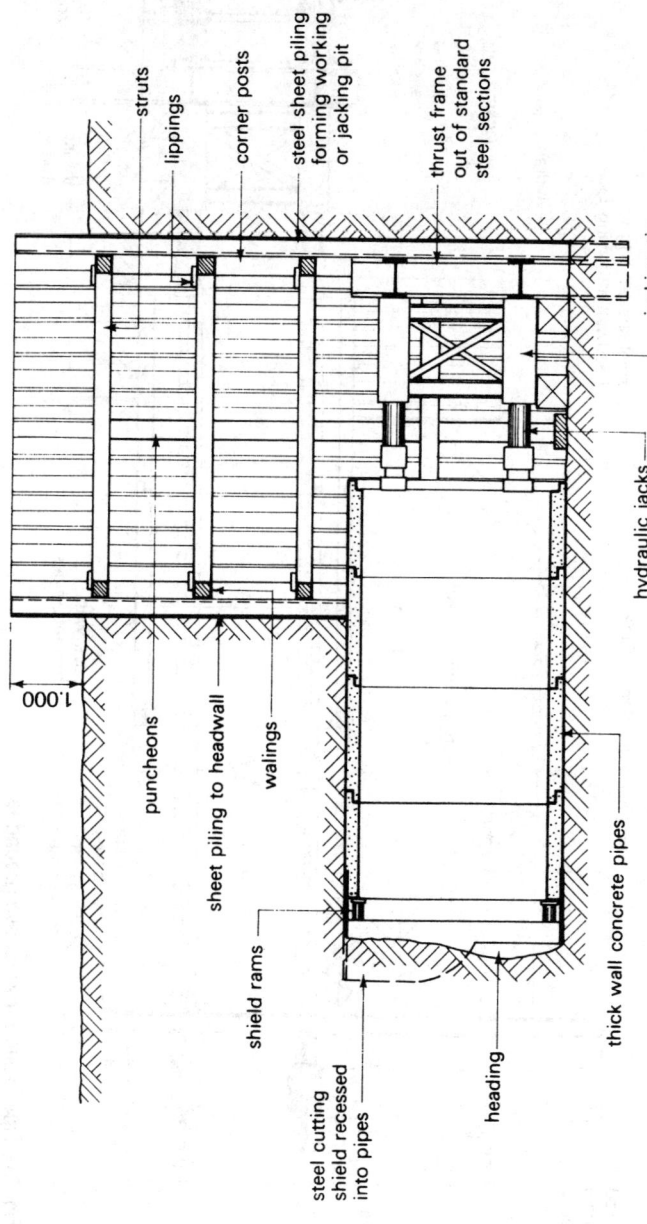

Fig. I.10 Pipe jacking from below ground level

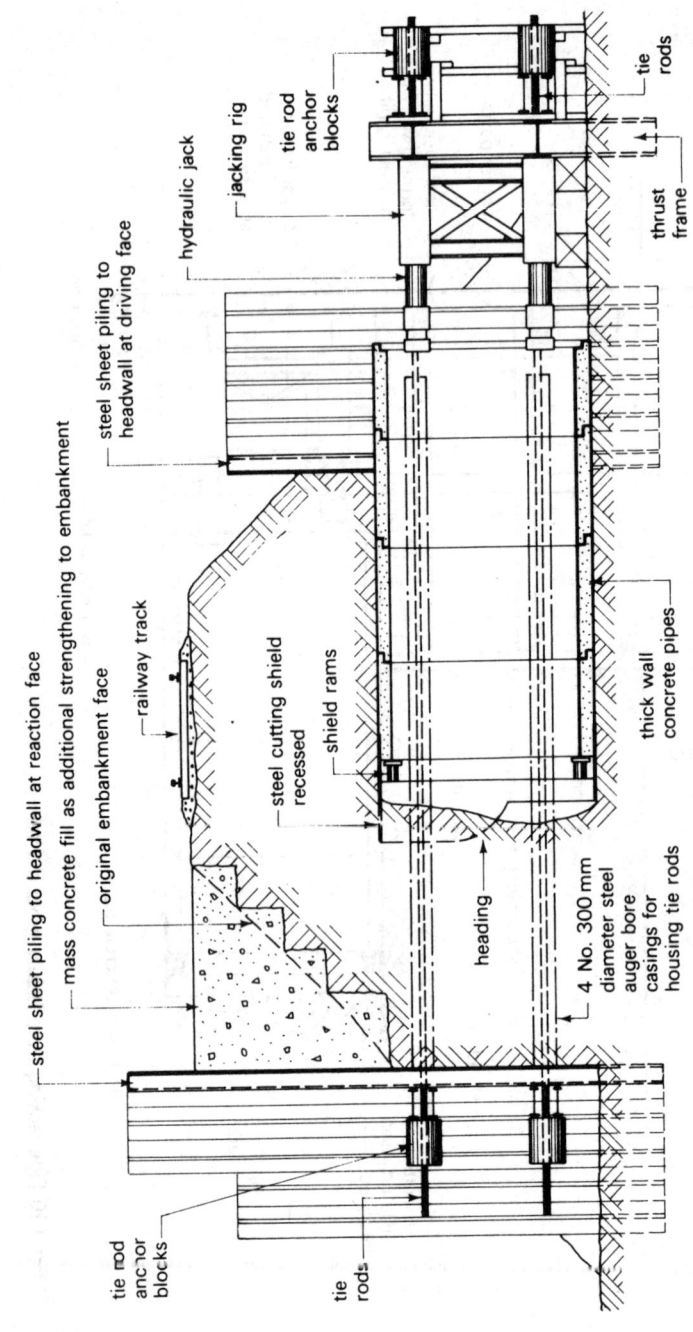

Fig. I.11 Pipe jacking from above ground level

driving process. This is a very safe method since the excavation work is carried out from within the casing or liner and the danger of collapsing excavations is eliminated; there is also no disruption of the surface or underground services and it is a practical method for most types of subsoil.

The most common method is to work from a jacking or working pit which is formed in a similar manner to traditional shafts except that a framed thrust pad is needed from which to operate the hydraulic jacks. The working pit must be large enough for the jacks to be extended and to allow for new pipe sections to be lowered into the working bay at the bottom — see Fig. 1.10.

Pipe jacking can also be carried out from ground level and is particularly suitable for driving pipes through an embankment to form a pedestrian subway. A series of 300 mm diameter lined augered bore holes are driven through the embankment to accommodate tie bars which are anchored to a bulkhead frame on the opposite side of the embankment. The reactions from the ramming jacks are thus transferred through the tie bars to the bulkhead frame and the driving action becomes one of pushing and pulling — see Fig. I.11. In firm soils the rate of bore by this method is approximately 3.000 m per day.

Pipes can also be jacked, from ground level, into the earth at a gradient of up to 1:12 using a jacking block attached to a row of tension piles sited below the commencing level.

PIPES

The pipes used in the above techniques are usually classified in diameter ranges thus:

1. *Small pipes* — 150 to 900 mm diameter — thrust or auger bored.
2. *Medium pipes* — 900 to 1 800 mm diameter — pipe jacking techniques.
3. *Large pipes* — 1 800 to 3 600 mm diameter — pipe jacking techniques.

Two materials are in common use for the pipes, namely concrete and steel. Spun concrete pipes are specially designed with thick walls and have a rubber joint making them especially suitable for sewers without the need for extra strengthening. Larger diameter pipes for pedestrian subway constructions are usually made of cast concrete and can have special bolted connections making the joints watertight which also renders them suitable for use as sewer pipes. Steel pipes have a wall thickness relative to their diameter and usually have welded joints to give high tensile strength, the alternative being a flanged and bolted joint. They are obtainable with various coatings and linings to meet special requirements such as corrosive-bearing effluents.

3
Demolition

This is a skilled and sometimes dangerous operation and unless of a very small nature should be entrusted to a specialist contractor. Demolition of a building or structure can be considered under two headings:

1. *Taking down* — partial demolition of a structure.
2. *Demolition* — complete removal of a structure.

Before any taking down or demolition is commenced it is usual to remove carefully all saleable items such as copper, lead, steel fittings, domestic fittings, windows, doors and frames.

Taking down requires a good comprehensive knowledge of building construction and design so that load bearing members and walls can be correctly identified and adequately supported by struts, props and suitable shoring. Most partial demolition works will need to be carried out manually using hand tools such as picks and hammers.

SURVEY

Before any works of demolition are started a detailed survey and examination of the building or structure and its curtilage should be made. Photographs of any existing defects on adjacent properties should be taken, witnessed and stored in a safe place. The relationship as well as the condition of adjoining properties which may be affected by the demolition should also be considered and noted, taking into account the existence of easements, wayleaves, party rights and boundary walls.

Roofs and framed structures: check whether proposed order of demolition will cause unbalanced thrusts to occur.

Walls: check whether these are load bearing, party or crosswalls. Examine condition and thickness of walls to be demolished and those to be retained.

Basements: careful examination required to determine if these extend under public footpaths or beyond boundary of site.

Cantilevers: check nature of support to balconies, heavy cornices and stairs.

Services these may have to be sealed off, protected or removed and could include any or all of the following:

1. Drainage runs.
2. Electricity cables.
3. Gas mains and service pipes.
4. Water mains and service pipes.
5. Telephone cables above and below ground level.
6. Radio and television relay cables.
7. District heating mains.

A careful survey of the whole site is advisable to ensure that any flammable or explosive materials such as oil drums and gas cylinders are removed before the demolition work commences. If the method of construction of the existing structure is at all uncertain all available drawings should be carefully studied and analysed or alternatively a detailed survey of the building should be conducted under the guidance of an experienced surveyor.

Adequate insurance should be taken out by the contractor to cover all claims from workmen, any third party and claims for loss or damage to property including roads, pavings and services.

STATUTORY NOTICES

Under the Public Health Act 1961 the local authority for the area in England or Wales must be notified before any demolition work can be started by the building owner or his agent. In the inner London area the notification is made to the district surveyor and in Scotland a warrant is required from the building authority of the burgh or county in which the proposed demolition work is to take place. It is also necessary in Great Britain to inform HM District Construction Inspector of Factories of any demolition works likely to take more than 6 weeks before such demolition commences.

Prior to the commencement of any demolition work the building owner or his agent must notify the Gas, Electricity and Water Area Boards, the Post Office telephone or telecommunication service and the companies responsible for other installations such as radio and television relay lines. It must be noted that it is the contractor's responsibility to ensure that all the services and other installations have been rendered safe or removed by the authority or company concerned.

SUPERVISION AND SAFETY

Part X of The Construction (General Provisions) Regulations 1961 covers all aspects of supervision and safety in respect of demolition works. The demolition contractor is required, under these regulations, to appoint a competent person to supervise the work. He must be experienced in the type of demolition work concerned and where more than one contractor is involved each must appoint a competent supervisor.

METHODS OF DEMOLITION

There are several methods of demolition and the choice is usually determined by:

1. *Type of structure* — for example, 2-storey framed structure, reinforced concrete chimney.
2. *Type of construction* — such as masonry wall, prestressed concrete, structural steelwork.
3. *Location of site* — a detached building on an isolated site which is defined as a building on a site where the minimum distance to the boundary is greater than twice the height of the building to be demolished. A confined site is where not all the boundaries are at a distance exceeding twice the height of the building to be demolished.

Hand demolition: involves the progressive demolition of a structure by operatives using hand-held tools; lifting appliances may be used to hoist and lower members or materials once they have been released. Buildings are usually demolished, by this method, in the reverse order to that of their construction storey by storey. Debris should only be allowed to fall freely where the horizontal distance from the point of fall to the public highway or an adjoining property is greater than 6.000 m or half the height from which the debris is dropped whichever is the greater. In all other cases a chute or skip should be used.

Pusher arm demolition: a method of progressive demolition using a machine fitted with a steel pusher arm exerting a horizontal thrust on to the building fabric. This method should only be used when the machine can be operated from a firm level base with a clear operating base of at least 6.000 m. The height of the building should be reduced by hand demolition if necessary to ensure that the height above the pusher arm does not exceed 600 mm. The pusher arm should not be overloaded and generally should be operated from outside the building. An experienced operator is required and he should work from within a robust cab capable of withstanding the impact of flying debris and be fitted with shatter proof glass cab windows. Where this method of demolition is adopted in connection with attached buildings, the structure to be demolished should first be detached from the adjoining structure by hand demolition techniques.

Deliberate collapse demolition: involves the removal of key structural members causing complete collapse of the whole or part of the building. Expert engineering advice should be obtained before this method is used. It should only be used on detached isolated buildings on reasonably level sites so that the safety to personnel can be carefully controlled.

Demolition ball techniques: a method of progressive demolition carried out by swinging a weight or demolition ball, suspended from a lifting appliance such as a crane, against the fabric of the structure. Three techniques can be used:

1. Vertical drop.
2. Swinging in line with the jib.
3. Slewing jib.

Whichever method is used a skilled operator is essential.

The use of a demolition ball from a normal duty mobile crane should be confined to the vertical drop technique only. A heavy duty machine such as a convertible dragline excavator should be used for the other techniques but in all cases an anti-spin device should be attached to the hoist rope. It is advisable to reduce the length of the crane jib as the demolition work proceeds but at no time should the jib head be less than 3.000 m above the part of the building being demolished.

Pitched roofs should be removed by hand demolition down to wall plate level and at least 50% to 70% of the internal flooring removed to allow for the free fall of debris within the building enclosure. Demolition should then proceed progressively storey by storey.

Demolition ball techniques should not be used on buildings over 30.000 m high since the fall of debris is uncontrollable. Attached buildings should be separated from the adjoining structure by hand demolition to leave a space of at least 6.000 m or half the height of the building, whichever is the greater; a similar clear space is required around the perimeter of the building to give the machine operating space.

Wire rope pulling demolition: only steel wire ropes should be used and the size should be adequate for the purpose but in no case less than 38 mm circumference. The rope should be firmly attached at both ends and the pulling tension gradually applied. No person should be forward of the winch or on either side of the rope within a distance of three quarters of the length between the winch and the building being demolished.

If after several pulls the method does not cause collapse of the structure it may have been weakened and should not therefore be approached but demolished by an alternative means such as a demolition ball or a pusher arm. A well-anchored winch or heavy vehicle should be used to apply the pulling force, great care being taken to ensure that the winch or the vehicle does not lift off its mounting, wheels or tracks.

Demolition by explosives: this is a specialist method where charges of explosives are placed within the fabric of the structure and detonated to cause partial or complete collapse. It should never be attempted by a building contractor without the advice and supervision of an expert.

Other methods: where site conditions are not suitable for the use of explosives the following specialist's methods can be considered:

1. *Gas expansion burster* — a steel cylinder containing a liquified gas, which expands with great force when subjected to an electric charge, is inserted into a prepared cavity in the fabric to be demolished. On being fired the expansion of the cylinder causes the fabric mass to be broken into fragments.

2. *Hydraulic burster* — consists of a steel cylinder with a number of pistons which are forced out radially under hydraulic pressure.

3. *Thermal reaction* — a structural steel member which is to be cut out and removed is surrounded by a mixture of a metal oxide and a reducing agent. This covering is ignited, usually by an electric current, which results in a liberation of a large quantity of heat causing the steel to become plastic. A small force such as a wire rope attached to a winch will be sufficient to cause collapse of the member.

4. *Thermic lance* — a steel tube, sometimes packed with steel rods, through which oxygen is passed. The tip of the lance is preheated by conventional means to melting point (approximately 1 000°C) when the supply of oxygen is introduced. This sets up a thermo-chemical reaction giving a temperature of around 3 500°C at the reaction end

which will melt all the materials normally encountered causing very little damage to surrounding materials.

The dangers and risks encountered with any demolition works cannot be over-emphasised and all builders should seek the advice of and employ specialist contractors to carry out all but the simple demolition tasks.

4 Underpinning

Part II Foundations

The main objective of underpinning is to transfer the load carried by an existing foundation from its present bearing level to a new level at a lower depth. It can also be used to replace an existing weak foundation.

Underpinning may be necessary for one or more of the following reasons:

1. (a) uneven loading;
   (b) unequal resistance of the subsoil;
   (c) action of tree roots;
   (d) action of subsoil water;
   (e) cohesive soil settlement.
2. To increase the load-bearing capacity of a foundation, which may be required to enable an extra storey to be added to the existing structure or if a change of use would increase the imposed loadings.
3. As a preliminary operation to lowering the adjacent ground level when constructing a basement at a lower level than the existing foundations of the adjoining structure or when laying deep services near to or below the existing foundations.

SITE SURVEY AND PRELIMINARY WORKS

Before any underpinning is commenced the following surveying and preliminary work should be carried out:

1. Notice should be served to the adjoining owners setting out in detail the intention to proceed with the proposed works and giving full details of any proposed temporary supports such as shoring.
2. A detailed survey of the building to be underpinned should be made recording any defects, cracks, supplemented by photographs and agreed with the building owner where possible.
3. Glass slips or 'tell tales' should be fixed across any vertical and lateral cracks to give a visual indicator of any movement taking place.
4. A series of check levels should be taken against a reliable datum or alternatively metal studs can be fixed to the external wall and their levels noted. These levels should be checked periodically as the work proceeds to enable any movements to be recorded and the necessary remedial action taken.
5. Permission should be obtained, from the adjoining owner, to stop up all flues and fireplaces to prevent the nuisance and damage which can be caused by falling soot.
6. If underpinning is required to counteract unacceptable settlement of the existing foundations an investigation of the subsoil should be carried out to determine the cause and to forecast any future movement so that the underpinning design will be adequate.
7. The loading on the structure should be reduced as much as possible by removing imposed floor loads and installing any shoring that may be necessary.

WALL UNDERPINNING

Traditional underpinning to walls is carried out by excavating in stages alongside and underneath the existing foundation, casting a new foundation, building up to the underside of the existing foundation in brickwork or concrete and finally pinning between the old and new work with a rich dry mortar.

To prevent the dangers of fracture or settlement the underpinning stages or bays should be kept short and formed to a definite sequence pattern so that no two bays are worked consecutively. This will enable the existing foundation and wall to arch or span the void created underneath prior to underpinning. The number and length of the bays will depend upon the following factors:

1. Total length of wall to be underpinned.
2. Width of existing foundation.
3. General condition of existing substructure.
4. Superimposed loading of existing foundation.

5. Estimated spanning ability of existing foundation.
6. Subsoil conditions encountered.

The generally specified maximum length for bays used in the underpinning of traditional wall construction is 1.500 m with the proviso that at no time should the sum total of unsupported lengths exceed 25% of the total wall length.

Bays are excavated and timbered as necessary after which the bottom of the excavation is prepared to receive the new foundation. To give the new foundation strip continuity, dowel bars are inserted at the end of each bay. Brick underpinning is toothed at each end to enable the bonding to be continuous whereas concrete underpinning usually has splice bars or dowels projecting to provide the continuity. Brickwork would normally be in a class 'B' brick bedded in 1:3 cement mortar laid in English bond for strength. Concrete used in underpinning is usually specified as a 1:2:4/20 mm aggregate mix using rapid hardening cement. The final pinning mix should consist of 1 part rapid hardening cement to 3 parts of well graded fine aggregate from 10 mm down to fine sand with a water/cement ratio of 0.35. In both methods the projection of the existing foundation is cut back to the external wall line so that the loads are transmitted to the new foundation and not partially dissipated through the original foundation strip on to the backfill material – see Fig. II.1.

PRETEST METHOD OF UNDERPINNING

This method is designed to prevent further settlement of foundations after underpinning has been carried out by consolidating the soil under the new foundation before the load from the underpinning is applied. The perimeter of the wall to be underpinned is excavated in stages as described for wall underpinning, the new foundation strip is laid and a hydraulic jack supporting a short beam is placed in the centre of the bay under the existing foundation. A dry mortar mix is laid between the top of the beam and the existing foundation and before it has finally set the jack is extended to give a predetermined load on the new foundation, thus pretesting the soil beneath.

This process is repeated along the entire underpinning length until the whole wall is being supported by the hydraulic jacks. Underpinning is carried out using brickwork or concrete walling between the jacks which are later removed and replaced with underpinning to complete the operation.

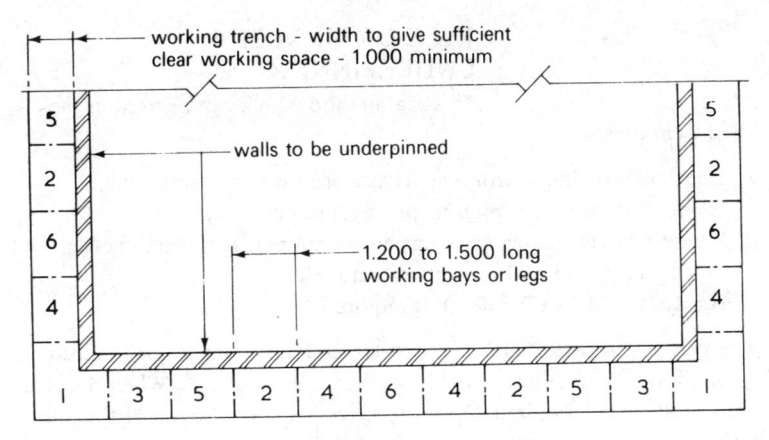

Typical underpinning schedule

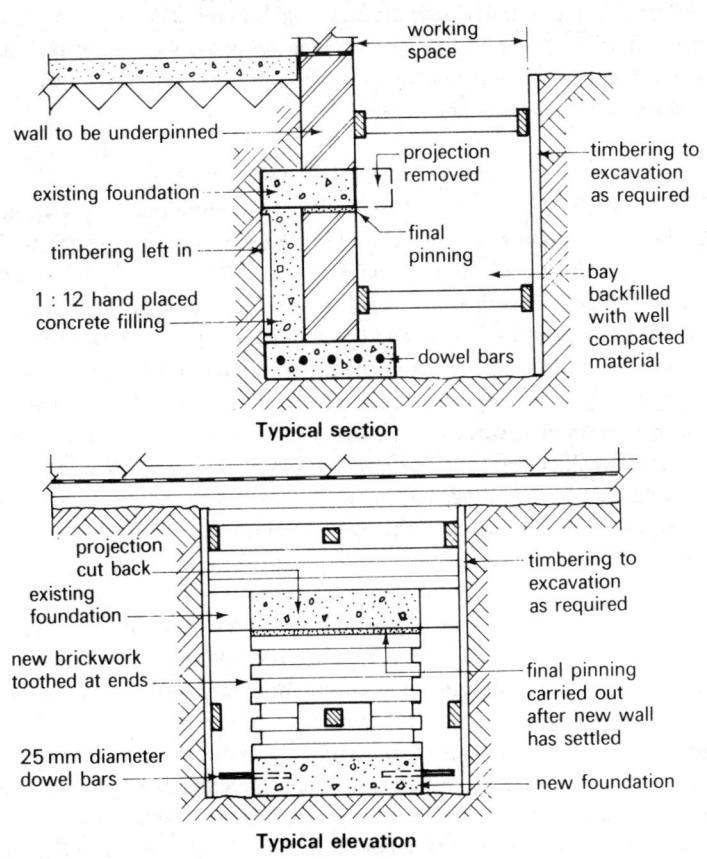

Typical section

- wall to be underpinned
- existing foundation
- timbering left in
- 1 : 12 hand placed concrete filling
- working space
- projection removed
- final pinning
- dowel bars
- timbering to excavation as required
- bay backfilled with well compacted material
- working trench - width to give sufficient clear working space - 1.000 minimum
- walls to be underpinned
- 1.200 to 1.500 long working bays or legs

Typical elevation

- projection cut back
- existing foundation
- new brickwork toothed at ends
- 25 mm diameter dowel bars
- timbering to excavation as required
- final pinning carried out after new wall has settled
- new foundation

Fig. II.1 Traditional brick underpinning

33

JACK OR MIGA PILE UNDERPINNING

This is a method which can be used in the following circumstances:

1. Depth of suitable bearing capacity subsoil is too deep to make traditional wall underpinning practical or economic.
2. Where a system giving no vibration is required. It is worth noting that this method is also practically noiseless.
3. If a system of flexible depth is required.

The system consists of short precast concrete pile lengths jacked into the ground until a suitable subsoil is reached — see Fig. II.2. When the jack pile has reached the required depth the space between the top of the pile and underside of the existing foundation is filled with a pinned concrete cap. The existing foundation must be in a good condition since in the final context it will act as a beam spanning over the piles. The condition and hence the spanning ability of the existing strip foundation will also determine the spacing of the piles.

NEEDLE AND PILES

If the wall to be underpinned has a weak foundation that is considered unsuitable for spanning over the heads of jack piles an alternative method giving the same degree of flexibility can be used. This method uses pairs of jacks or usually bored piles in conjunction with an *in situ* reinforced concrete beam or needle placed above the existing foundation. The system works on the same principle as a dead shoring arrangement relying on the arching effect of bonded brickwork. If water is encountered when using bored piles a pressure pile can be used as an alternative. The formation of both types of pile is described in the following chapter on piling. Typical arrangements to enable the work to be carried out from both sides of the wall or from the external face only are shown in Fig. II.3.

Pynford Stooling Method

The Pynford method of underpinning enables walls to be underpinned in continuous runs without the use of needles or raking shoring. The procedure is to cut away portions of brickwork, above the existing foundation, to enable precast concrete or steel stools to be inserted and pinned. The intervening brickwork can be removed, leaving the structure entirely supported on the stools. Reinforcing bars are threaded between and around the stools and caged to form the ring beam reinforcement. After the formwork has been placed

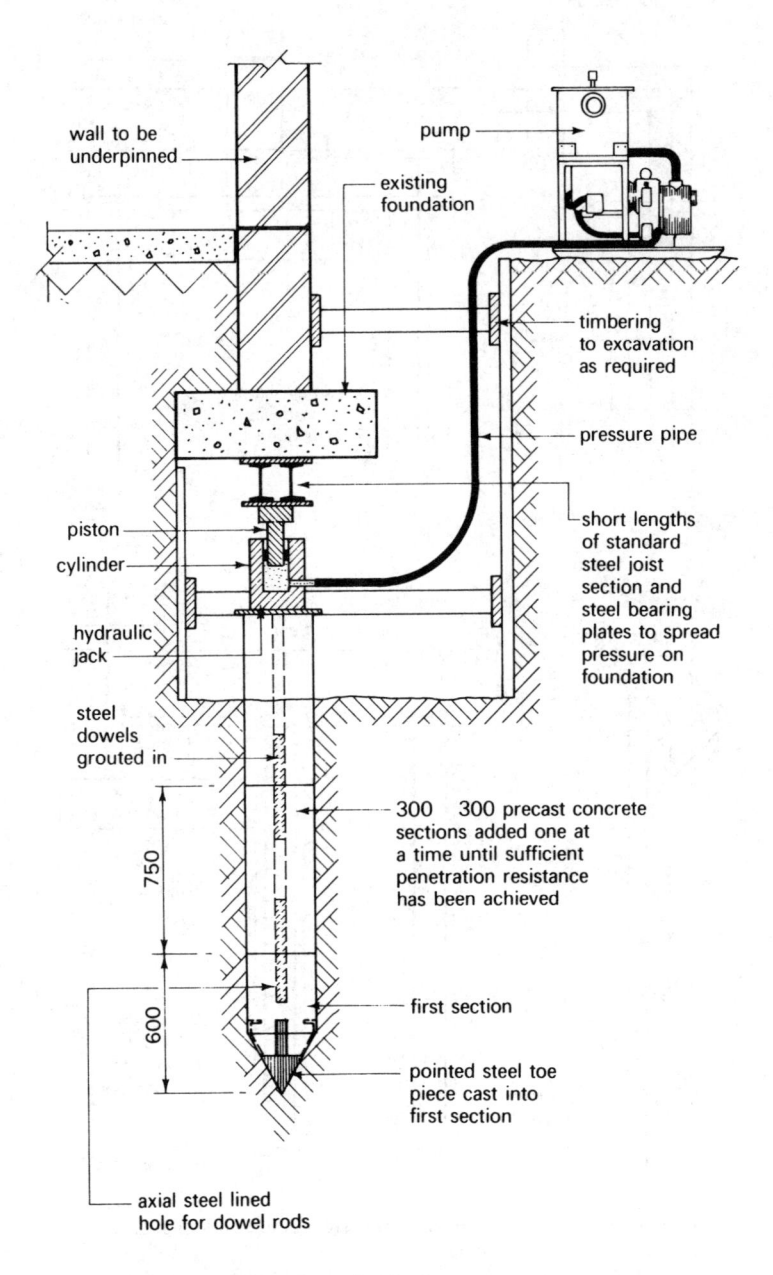

Fig. II.2 Jack or miga pile underpinning

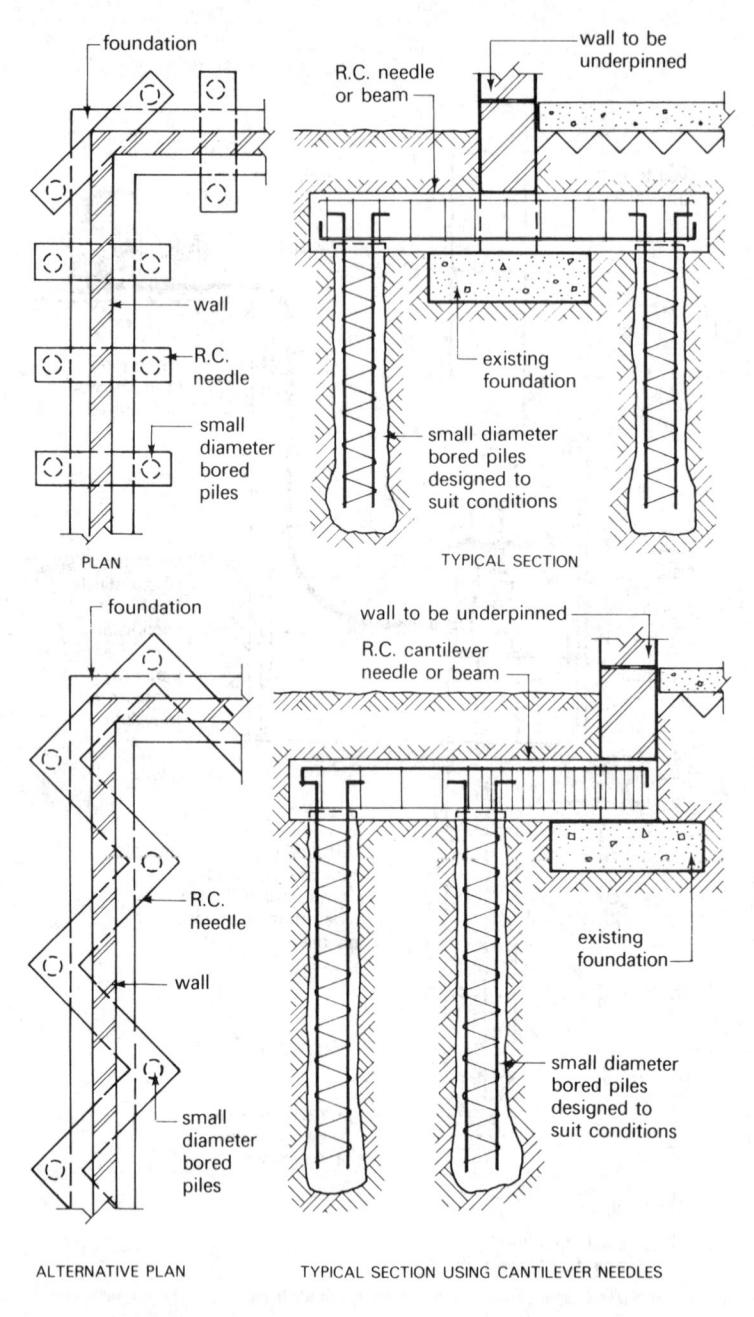

Fig. II.3 Needle and pile underpinning

and the beam cast, final pinning can be carried out using a well rammed dry mortar mix — see Fig. II.4. This method replaces the existing foundation strip with a reinforced concrete ring beam from which other forms of deep underpinning can be carried out if necessary.

Other forms of underpinning using beams

Stressed Steel: This consists of standard universal beam sections in short lengths of 600 to 1 500 mm long with a steel diaphragm plate welded to each end which is drilled to take high tensile torque bolts. Short lengths of wall are removed and the steel beam inserted. The joints between adjacent diaphragm plates are formed so that a small space occurs on the lower or tension side to give a predetermined camber when the bolts are tightened. Final pinning between the top of the stressed steel beam and the wall completes the operation.

Prestressed concrete: short precast concrete blocks are inserted over the existing foundations as the brickwork is removed. The blocks are formed to allow for post tensioning stressing tendons to be inserted, stressed and anchored to form a continuous beam. Final pinning completes the underpinning.

It should be noted that all forms of beam underpinning can also be used to form a lintel or beam, in any part of a wall, prior to the formation of a large opening.

FINAL PINNING

Although final pinning is usually carried out by ramming a stiff dry cement mortar mix into the space between the new underpinning work and the existing structure, alternative methods are available such as:

1. *Flat jacks* — circular or rectangular hollow plates of various sizes made from thin sheet metal which can be inflated with high pressure water for temporary pinning or, if work is to be permanent, with a strong cement grout. The increase in thickness of the flat jacks is approximately 25 mm.
2. *Wedge bricks* — special bricks of engineering quality of standard face length but with a one brick width and a depth equal to two courses. The brick is made in two parts, the lower section having a wide sloping channel in its top bed surface to receive the wedge shaped and narrower top section. Both parts have a vertical slot through which the bedding grout passes to key the two sections together.

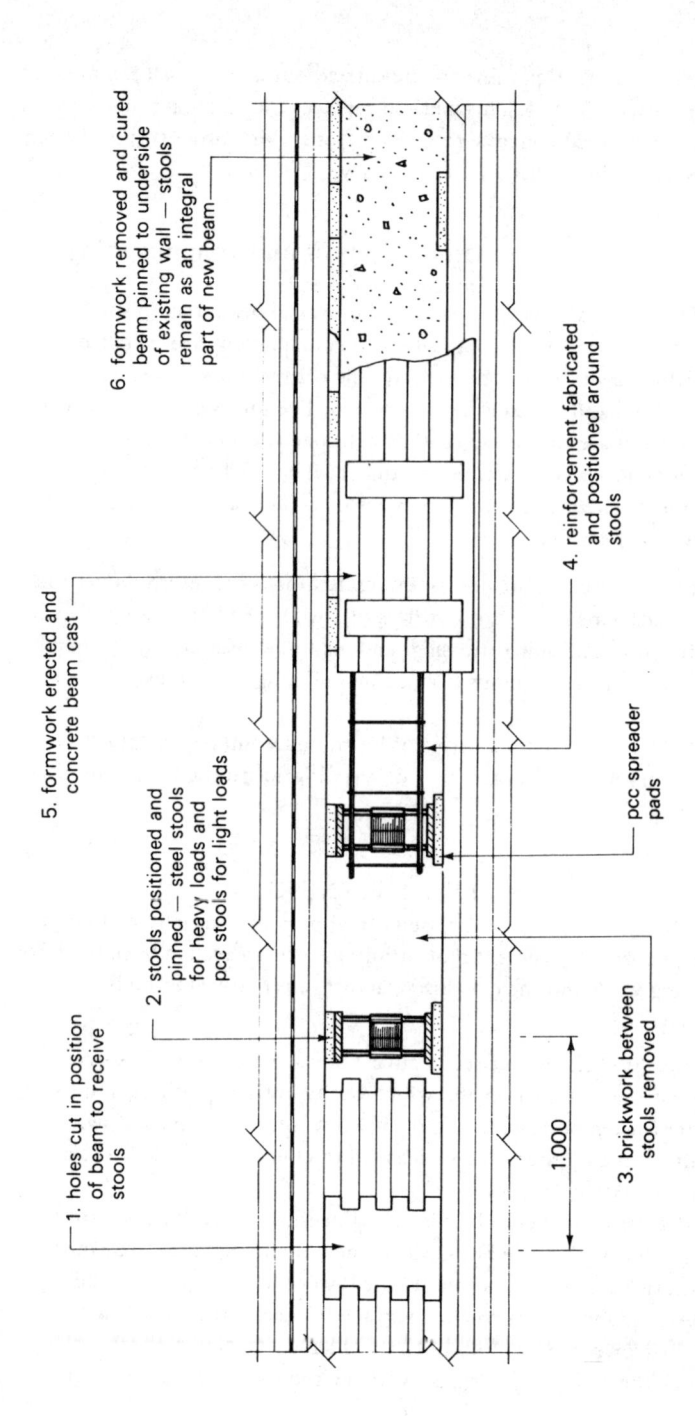

Fig. II.4 'Pynford' stooling method of underpinning

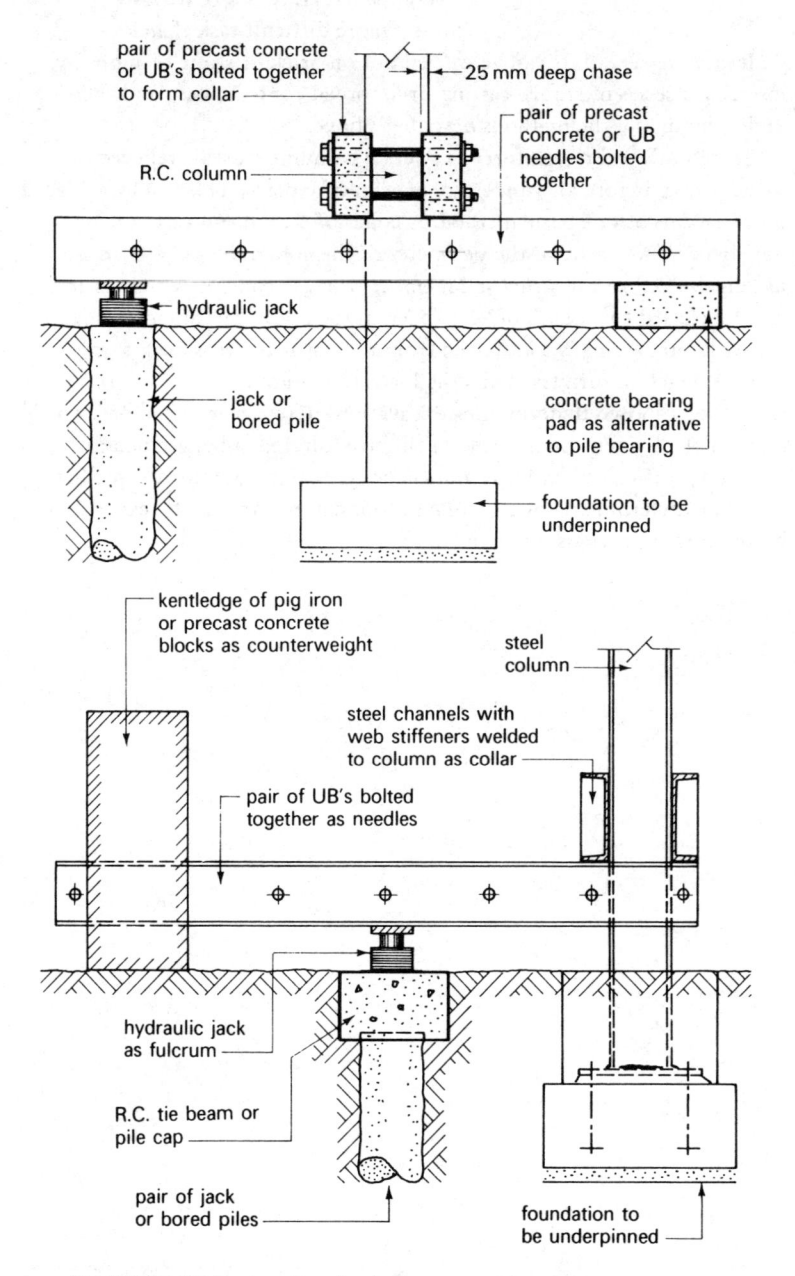

Fig. II.5 Typical column underpinning arrangements

UNDERPINNING COLUMNS

This is a more difficult task than underpinning a wall. It can be carried out on brick or stone columns by inserting a series of stools, casting a reinforced concrete base and then underpinning by the methods described above.

Structural steel or reinforced concrete columns must be relieved of their loading before any underpinning can take place. This can be achieved by variations of one basic method. A collar of steel or precast concrete members is fixed around the perimeter of the column. Steel collars are usually welded to the structural member whereas concrete columns are usually chased to a depth of 25 to 50 mm to receive the support collar. The column loading is transferred from the collar to cross beams or needles which in turn transmits the loads to the ground at a safe distance from the proposed underpinning excavations. Cantilever techniques which transfer the loadings to one side of the structural member are possible providing sufficient kentledge and anchorage can be obtained – see Fig. II.5. The underpinning of the column foundation can now be carried out by the means previously described.

5
Piled foundations

A pile can be loosely defined as a column inserted in the ground to transmit the structural loads to a lower level of subsoil. Piles have been used in this context for hundreds of years and until the twentieth century were invariably of driven timber. Today a wide variety of materials and methods are available to solve most of the problems encountered when confronted with the need for deep foundations. It must be remembered that piled foundations are not necessarily the answer to all awkward foundation problems but should be considered as an alternative to other techniques when suitable bearing capacity soil is not found near the lowest floor of the structure.

The unsuitability of the upper regions of a subsoil may be caused by:

1. Low bearing capacity of the subsoil.
2. Heavy point loads of the structure exceeding the soil bearing capacity.
3. Presence of highly compressible soils near the surface such as filled ground and underlying peat strata.
4. Subsoils such as clay which may be capable of moisture movement or plastic failure.
5. High water table.

CLASSIFICATION OF PILES

Piles may be classified by the way in which they transmit their loads to the subsoils or by the way they are formed. Piles may transmit their loads to a lower level by:

1. *End bearing* — the shafts of the piles act as columns carrying the loads through the overlaying weak subsoils to firm strata into which the pile toe has penetrated. This can be a rock strata or a layer of firm sand or gravel which has been compacted by the displacement and vibration encountered during the driving.
2. *Friction* — any foundation imposes on the ground a pressure which spreads out to form a bulb of pressure. If a suitable load bearing strata cannot be found at an acceptable level, particularly in stiff clay soils, it is possible to use a pile to carry this bulb of pressure to a lower level where a higher bearing capacity is found. The friction or floating pile is mainly supported by the adhesion or friction action of the soil around the perimeter of the pile shaft.

In most situations piles work on a combination of the two principles outlined above but the major design consideration identifies the pile class.

Piles may be preformed and driven thus displacing the soil through which they pass and are therefore classified as displacement piles. Alternatively the soil can be bored out and subsequently replaced by a pile shaft and such piles are classified as replacement piles.

DOWNDRAG

When piles are driven through weak soils such as alluvial or silty clay, peat layers and reclaimed land the long term settlement and consolidation of the ground can cause downdrag of the piles. This is in simple terms a transference of the downdrag load of consolidating soil on to a pile shaft and is called negative skin friction since the resistance to this transfer of load is the positive skin friction of the pile shaft. Downdrag loads have been known to equal the design bearing capacity of the pile. The anticipated negative skin friction can be calculated but such evaluations are beyond the scope of this text.

Three methods to counteract the effects of downdrag are possible:

1. Increase the number of piles being used to carry the structural loads to accommodate the anticipated downdrag loads. This method may necessitate the use of twice the number of piles required to carry only the structural loadings.
2. Increase the size of the piles designed for structural loads only; this can be expensive, since the cost of forming piles does not increase pro rata with its size.
3. To design for the structural loads only and prevent the transference of downdrag loads by coating the external face of the shaft with a slip layer, thus reducing the negative skin friction. A special bituminous compound has been developed to fulfil this task and is

applied in a 10 mm thick coat to that part of the shaft where negative skin friction can be expected. The toe of the pile is never coated so that the end bearing capacity is not reduced. Care must be taken to ensure that the coating is not damaged during transportation, storage and driving particularly in hot weather when a reflective coating of lime wash may be necessary.

DISPLACEMENT PILES

General term applied to piles which are driven, thus displacing the soil, and includes those piles which are preformed, partially preformed or are driven *in situ* piles.

Timber piles: usually square sawn hardwood or softwood in lengths up to 12.000 m with section sizes ranging from 225 × 225 mm to 600 × 600 mm. They are easy to handle and can be driven by percussion with the minimum of experience. Most timber piles are fitted with an iron or steel driving shoe and have an iron ring around the head to prevent splitting due to driving. Although not particularly common they are used in sea defences such as groynes and sometimes as guide piles for large trestles in conjunction with steel sheet piling. Load bearing capacities can be up to 350 kN per pile depending upon section size and/or species.

Precast concrete piles: used on medium to large contracts where soft soils overlying a firm strata are encountered and at least 100 piles will be required. Lengths up to 18.000 m with section sizes ranging from 250 × 250 mm to 450 × 450 mm carrying loadings of up to 1 000 kN are generally economical for the conditions mentioned above. The precast concrete driven pile has little frictional bearing strength since the driving operation moulds the cohesive soils around the shaft which reduces the positive frictional resistance.

Problems can be encountered when using this form of pile in urban areas due to:

1. Transporting the complete length of pile through narrow and/or congested streets.
2. The driving process, which is generally percussion, can set up unacceptable noise and/or vibrations.
3. Many urban sites are in themselves restricted or congested thus making it difficult to manoeuvre the long pile lengths around the site.

Preformed concrete piles: available as reinforced precast concrete or prestressed concrete piles, but due to the problems listed above and the site difficulties which can be experienced in splicing or lengthening preformed piles their use has diminished considerably in recent years in favour of precast piles formed in segments or the partially preformed types of pile. Typical examples of the segmental type are West's 'Hardrive' and 'Segmental' piles. The 'Hardrive' pile is composed of standard interchangeable precast concrete units of 10.000, 5.000 and 2.500 m lengths designed to carry working loads up to 800 kN. The pile lengths are locked together with four H-section pins located at the corners of an aligning sleeve — see Fig. II.6. The 'Segmental' pile is designed for lighter loading conditions of up to 300 kN and is formed by joining the 1.000 m long standard lengths together with spigot and socket joints up to a maximum length of 15.000 m. Special half length head units are available to reduce wastage to a minimum — see Fig. II.7.

Steel preformed piles: used mainly in conjunction with marine structures and where overlying soils are very weak. Two forms are encountered, the box pile, which is made from welded rolled steel sections, and the BS4 universal bearing pile, which has the conventional 'H' section. These piles are relatively light and therefore are easy to handle and drive. Splicing can be carried out by site welding to form piles up to 15.000 m long with load bearing capacities up to 600 kN. Consideration must always be given as to the need to apply a protective coating to the pile to guard against corrosion.

Composite piles: sometimes referred to as partially preformed piles and are formed by a method which combines the use of precast and *in situ* concrete or steel and *in situ* concrete. They are used mainly on medium to large contracts where the presence of running water or very loose soils would render the use of bored or preformed piles as unsuitable. Composite piles provide a pile of readily variable length made from easily transported short lengths. Typical examples are 'Prestcore', West's 'Shell' and cased piles.

The 'Prestcore' pile is a composite pile formed inside a lined bored hole and is strictly speaking a replacement pile and is therefore described in more detail in the following section on replacement piles.

The 'Shell' pile is however a driven or displacement pile consisting of a series of precast shells threaded on to a mandrel and top driven to the required set. After driving and removing the mandrel the hollow core can be inspected, a cage of reinforcement can be inserted and the void filled with *in situ* concrete. Lengths up to 60.000 m with bearing capacities within the range of 500 to 1 200 kN are possible with this method. The precast concrete shells are reinforced by a patent system using fibrillated polypropylene film as a substitute for the traditional welded steel fabric — see Fig. II.8.

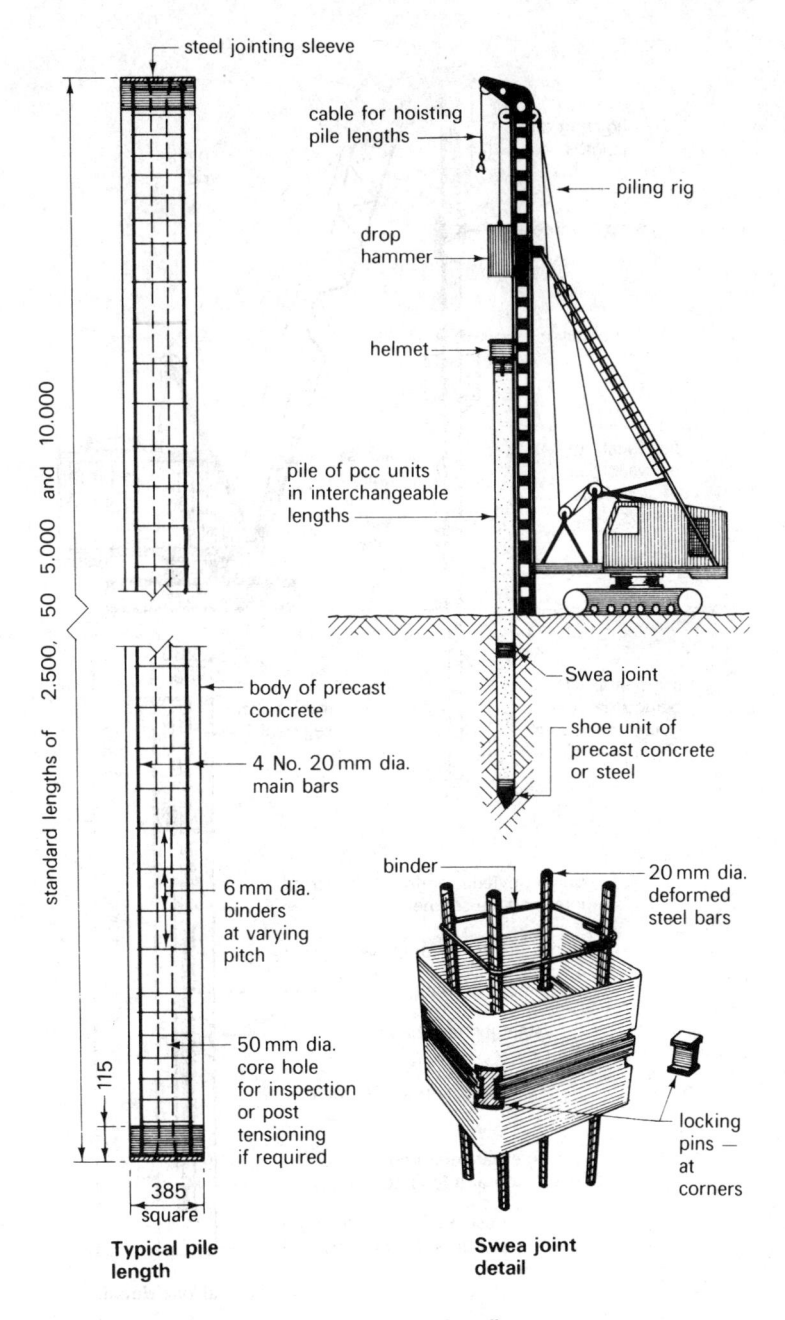

Fig. II.6 West's hardrive precast modular piles

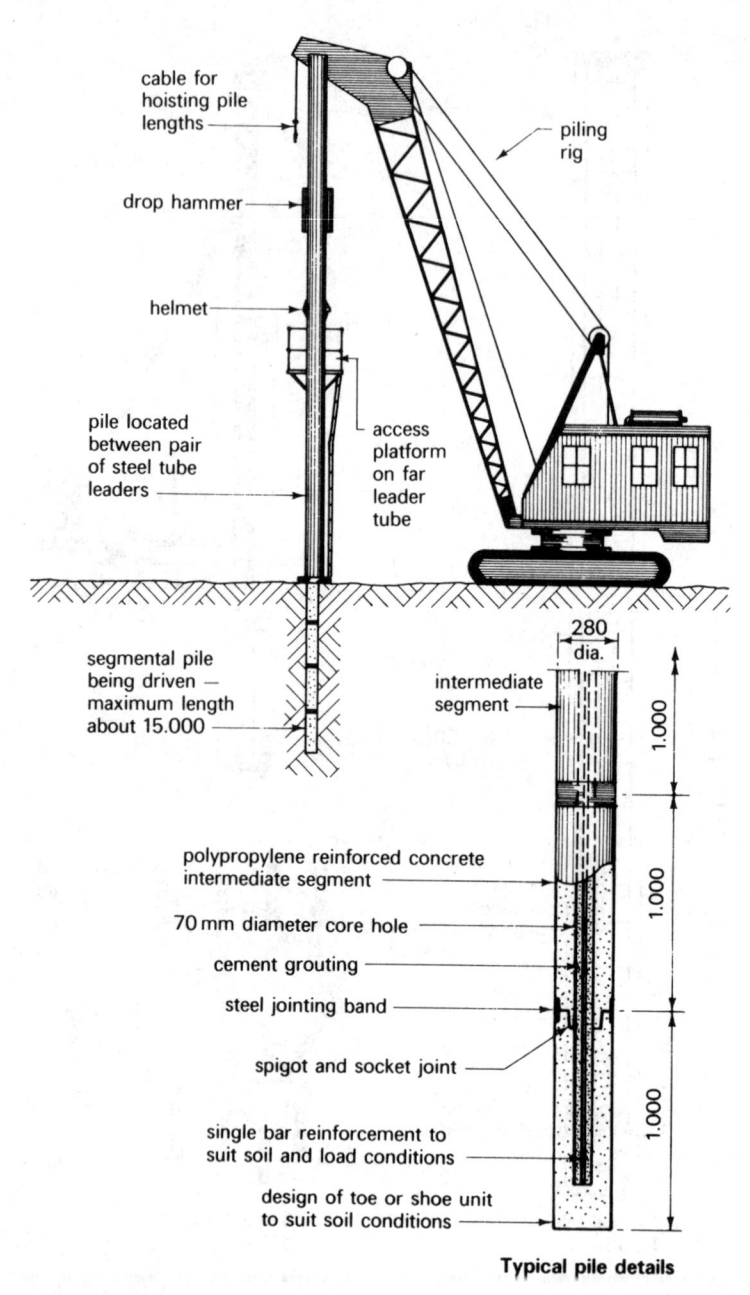

Fig. II.7 West's segmental piles

Typical pile details

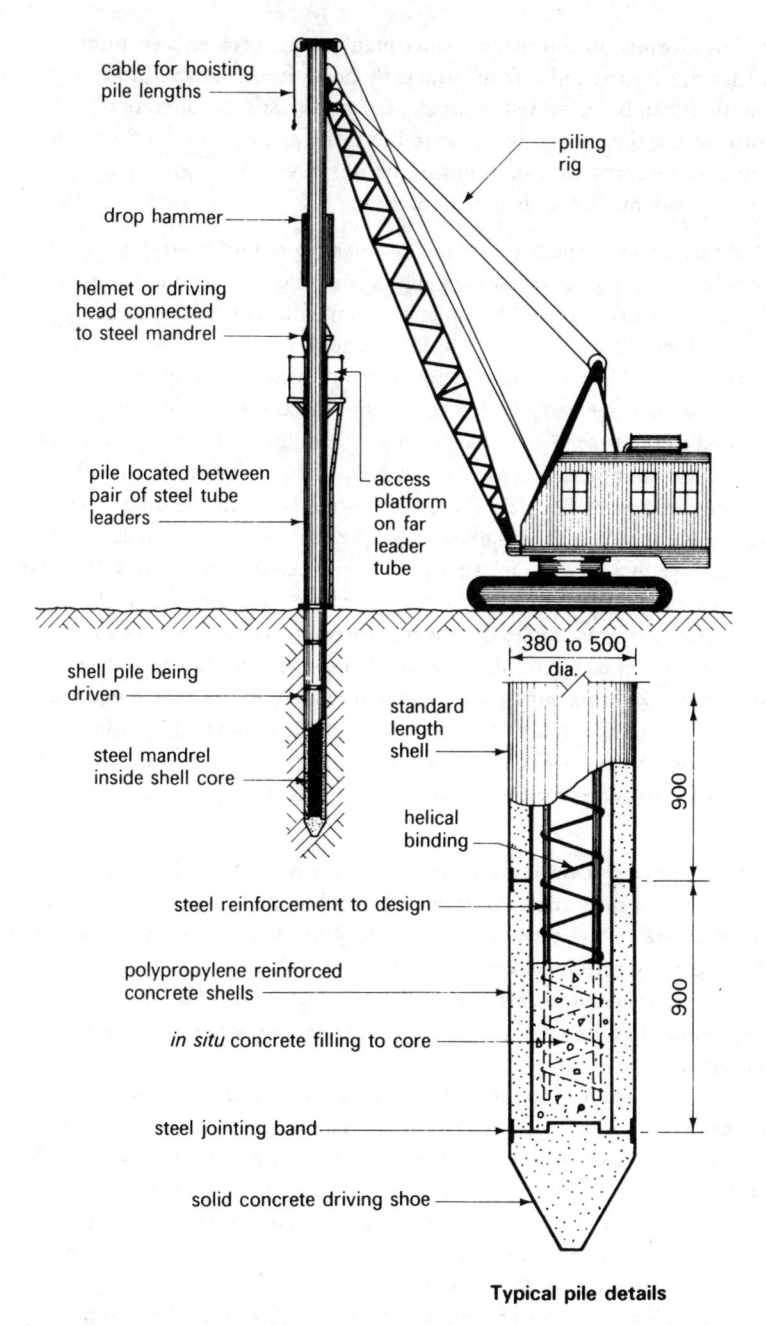

Fig. II.8 West's shell piles

Piles formed in this manner solve many of the problems encountered with waterlogged and soft substratas by being readily adaptable in length, the shaft can be inspected internally before the *in situ* concrete is introduced, the flow of water or soil into the pile is eliminated and the presence of corrosive conditions in the soil can be overcome by using special cements in the shell construction.

BSP cased piles: these are typical composite piles using steel and *in situ* concrete. Cased piles are bearing piles consisting of a driven tube which is filled with *in situ* concrete. The casing is manufactured from steel strip or plate which is formed into a continuous helix with the adjoining edges butt welded. They are usually driven into position by using an internal drop hammer operating within the casing. Usually a flat circular plate is welded to the base of the casing and a cushion plug of concrete with a very low water content is placed to a depth equal to 2½ times the pile diameter directly on top of the plate shoe. Pile lengths are available up to 24.000 m as a single tube but should extra length be required extension casings can be butt welded on after the first length has been driven to a suitable depth.

Apart from situations such as jetties, where a fair amount of the pile is left projecting, cased piles do not require reinforcement except for splice bars at the top to bond the pile to a pile cap. The *in situ* concrete is a standard 1:2:4 mix with a water/cement ratio of 0.4 to 0.5. A wide range of diameters from 250 to 600 mm are available with varying casing thicknesses to give working loads per pile ranging from 150 to 1 500 kN according to type of subsoil. Typical cased pile details are shown in Fig. II.9.

Driven *in situ* or cast in place piles: an alternative to preformed displacement piles and are suitable for medium and large contracts where there are likely to be variations in the lengths of piles required. They can be formed economically in diameters of 300 to 600 mm with lengths up to 18.000 m designed to carry loads cf up to 1 300 kN. They generally require heavy piling rigs, an open level site and a site where noise is unrestricted.

In some systems the tube which is used to form the pile shaft is top driven and in others such as the Franki system driving is carried out by means of an internal drop hammer working on a plug of dry concrete. *In situ* concrete for the core is introduced into the lined shaft through a hopper or skip and consolidation of the concrete can be carried out by impact of the internal drop hammer or by vibration of the tube as it is withdrawn — see Figs. II.10 and II.11.

A problem which can be encountered with this form of pile is necking due to ground water movement washing away some of the concrete thus

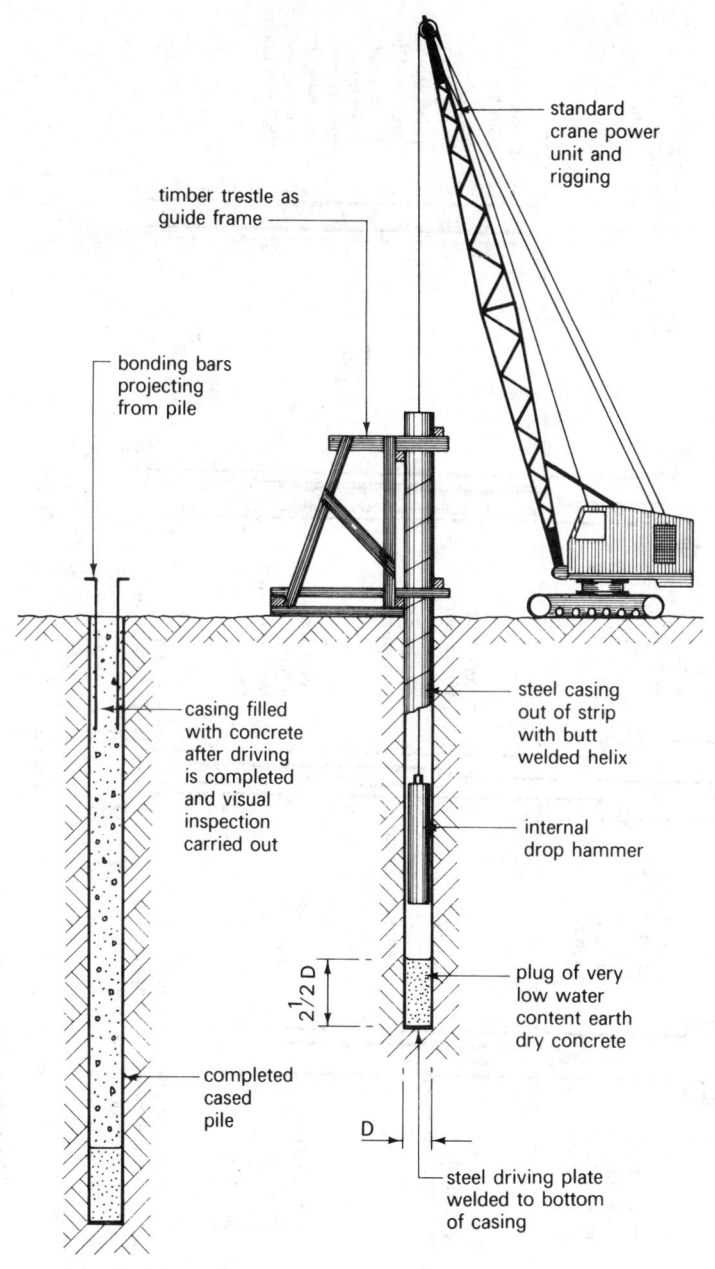

Fig. II.9 BSP cased piles

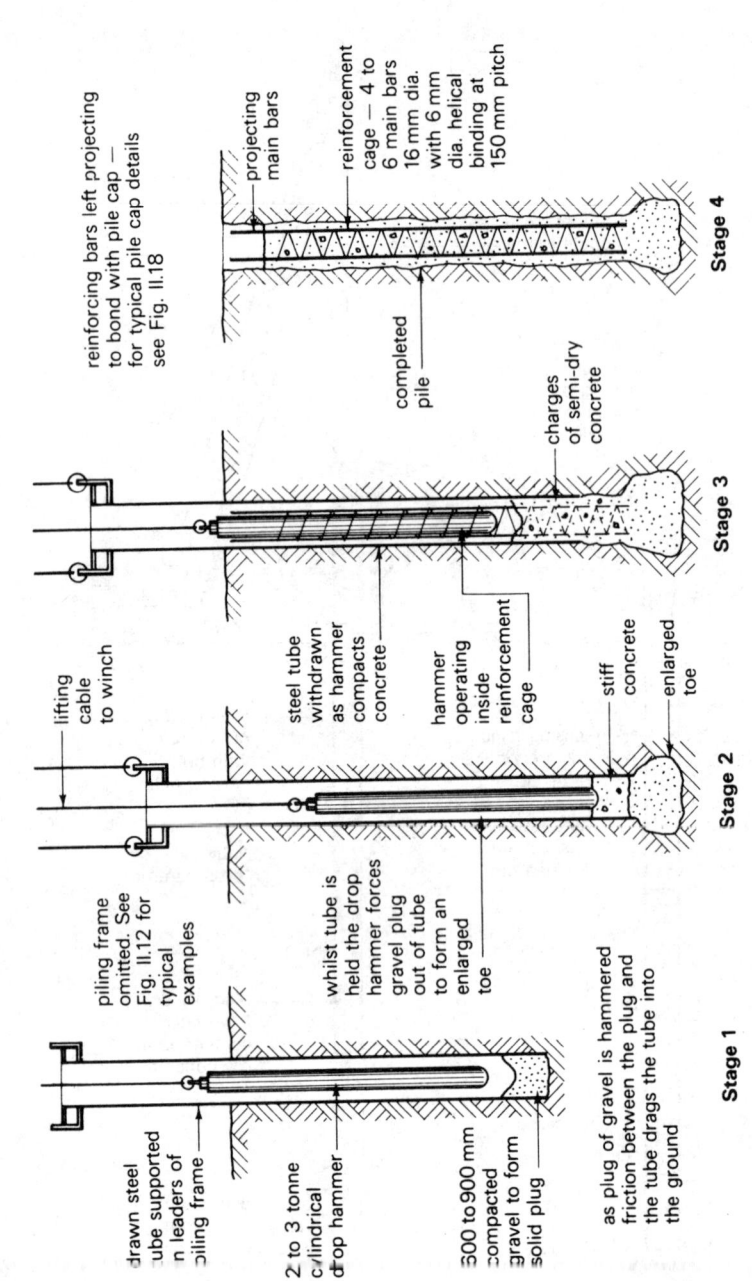

Fig. II.10 Franki driven *in situ* piles

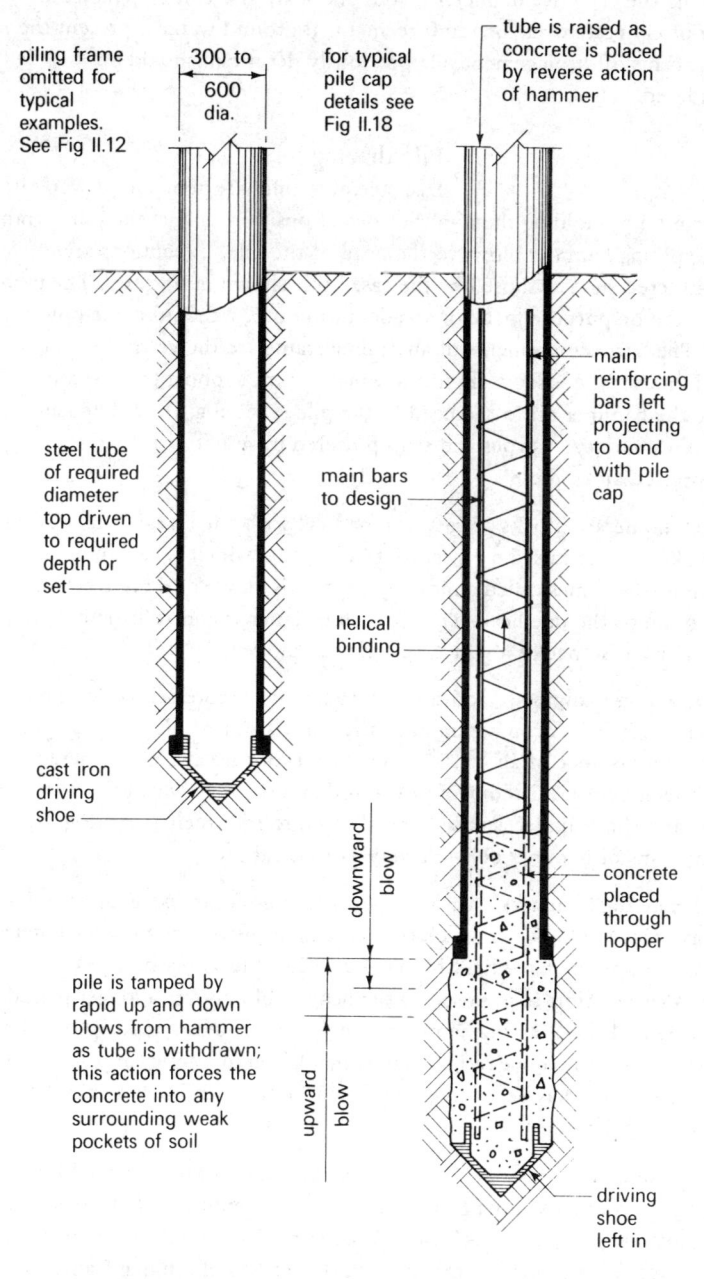

Fig. II.11 Vibro cast *in situ* pile

reducing the effective diameter of the pile shaft and consequently the cover of concrete over the reinforcement. If ground water is present the other forms of displacement pile previously described should be considered.

Pile driving

Displacement piles are generally driven into the ground by holding them in the correct position against the piling frame and applying hammer blows to the head of the pile. Exceptions are encountered such as those like the cased pile shown in Fig. II.9. The piling frame can be purpose made or an adaptation of a standard crane power unit. The basic components of any piling frame are the vertical member which houses the leaders or guides which in turn support the pile and guide the hammer on to the head of the pile — see Fig. II.12. Pile hammers come in a variety of types and sizes powered by gravity, steam, compressed air or diesel.

Drop hammers: blocks of cast iron or steel with a mass range of 1 500 to 8 000 kg and are raised by a cable attached to a winch. The hammer, which is sometimes called a monkey or ram, is allowed to fall freely by gravity on to the pile head. The free fall distance is controllable but generally a distance of about 1.200 m is employed.

Single acting hammers: activated by steam or compressed air have much the same effect as drop hammers in that the hammer falls freely by gravity through a distance of about 1.500 m. Two types are available wherein one case the hammer is lifted by a piston rod and those in which the piston is static and the cylinder is raised and allowed to fall freely — see Fig. II.12. Both forms of hammer deliver a very powerful blow.

Double acting hammers: activated by steam or compressed air consist of a heavy fixed cylinder in which there is a light piston or ram which delivers a large number of rapid light blows (90 to 225 blows per minute) in a short space of time as opposed to the heavier blows over a longer period of the drop and single acting hammers. The object is to try to keep the pile constantly on the move rather than being driven in a series of jerks. This type of hammer has been largely replaced by the diesel hammer and by vibration techniques.

Diesel hammers: designed to give a reliable and economic method of pile driving. Various sizes giving different energy outputs per blow are available but most deliver between 46 and 52 blows per minute. The hammer can be suspended from a crane or mounted in the leaders of a piling frame. A measured amount of liquid fuel is fed into a cup formed in the base of the

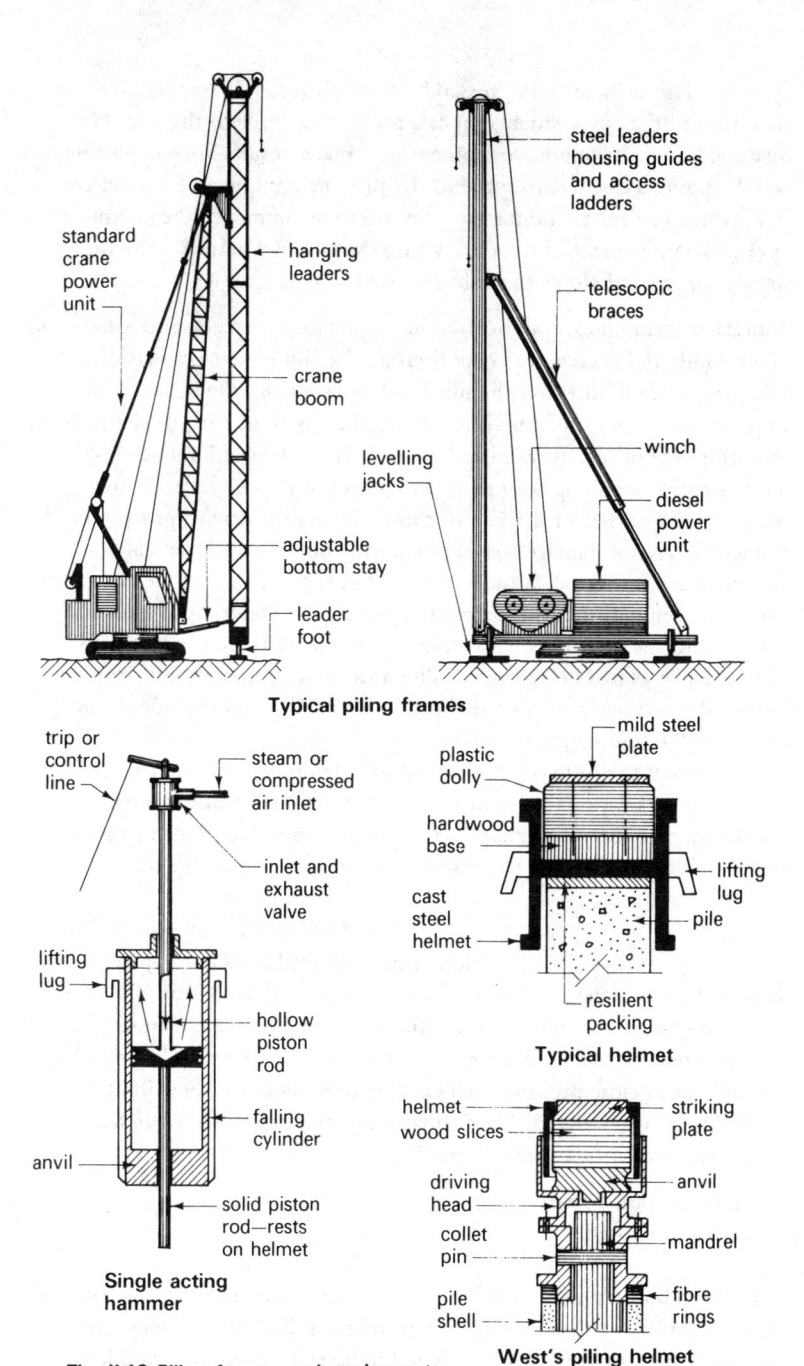

Fig. II.12 Piling frames and equipment

cylinder. The air being compressed by the falling ram is trapped between the ram and the anvil which applies a preloading force to the pile. The displaced fuel, at the precise moment of impact, results in an explosion which applies a downward force to the pile and an upward force on the ram, which returns to its starting position to recommence the complete cycle. The movement of the ram within the cylinder activates the fuel supply, opens and closes the exhaust ports — see Fig. II.13.

Vibration techniques: can be used in driving displacement piles where soft clays, sands and gravels are encountered. The equipment consists of a vibrating unit mounted on the pile head transmitting vibrations of the required frequency and amplitude down the length of the pile shaft. These vibrations are in turn transmitted to the surrounding soil, reducing its shear strength enabling the pile to sink into the subsoil under its own weight and also that of the vibrator unit. To aid the driving process and to reduce the risk of damage to the pile during driving, water jetting techniques can be used. Water is directed at the soil around the toe of the pile to loosen it and ease the driving process. The water pipes are usually attached to the perimeter of the pile shaft and are therefore taken down with the pile as it is being driven. The water jetting operation is stopped before the pile reaches its final depth so that the toe of the pile is finally embedded in undisturbed soil.

To protect the heads of preformed piles from damage due to impact from hammers various types of protective cushioned helmets are used. These can be either of a general nature as shown in Fig. II.12 or special purpose for a particular system such as those used for West's piling.

REPLACEMENT PILES

Sometimes referred to as bored piles and formed by removing a column of soil and replacing with *in situ* concrete or as in the case of composite piles with precast and *in situ* concrete. Replacement or bored piles are considered for sites where piling is being carried out in close proximity to existing buildings or where vibration and/or noise is restricted. The formation of this type of pile can be considered under two general classifications:

1. Percussion bored.
2. Rotary bored.

Percussion bored piles: suitable for small and medium sized contracts of up to 300 piles in both clay and/or gravel subsoils. Pile diameters are usually from 300 to 950 mm designed to carry loads up to 1 500 kN. Apart from the common factor with all replacement piles that the strata

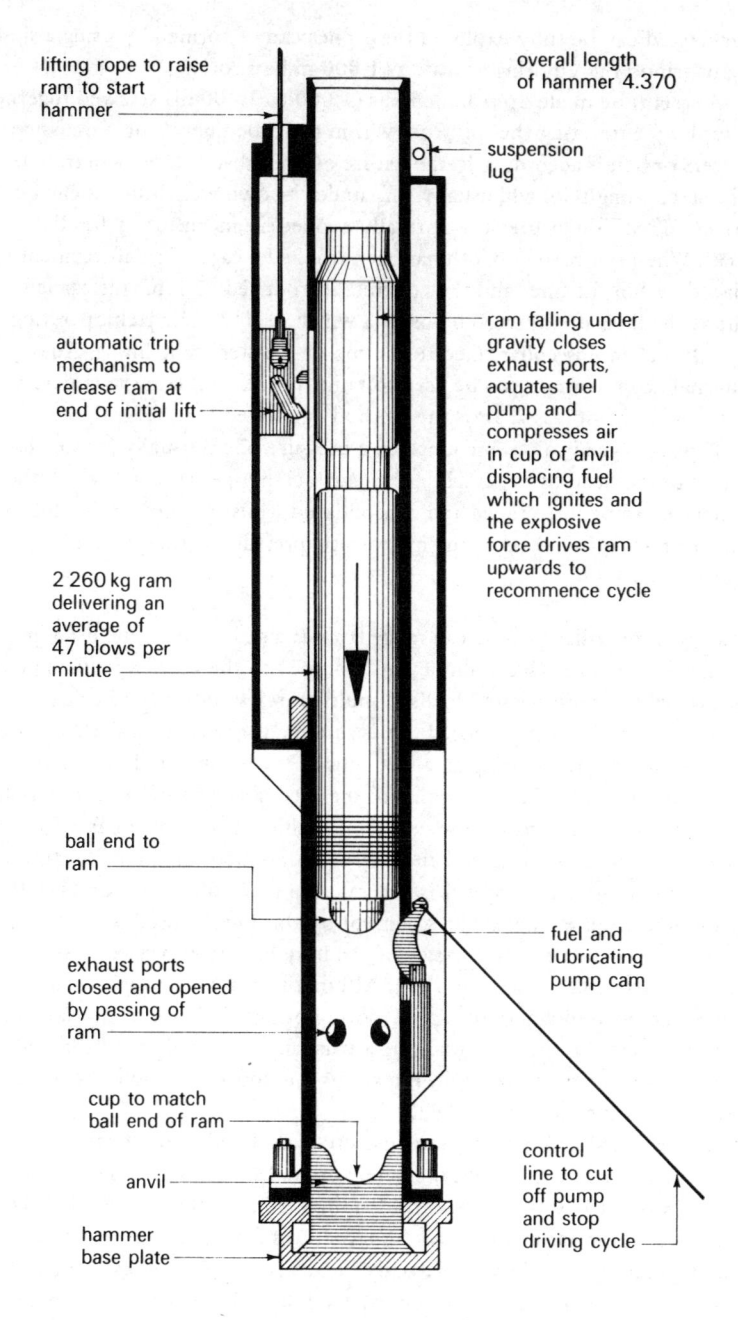

Fig II.13 Typical diesel hammer details

penetrated can be fully explored these piles can be formed by using a shear leg or tripod rig requiring as little as 1.800 m headroom.

A steel tube made up from lengths (1.000 to 1.400 m) screwed together is sunk by extracting the soil from within the tube liner using percussion cutters or balers according to the nature of the subsoil to be penetrated. The steel lining tube will usually sink under its own weight but it can be driven in with slight pressure normally applied by means of hydraulic jacks. When the correct depth has been reached a cage of reinforcement is placed within the liner and the concrete introduced. Tamping is carried out as the liner is extracted by using a winch or hydraulic jack operating against a clamping collar fixed to the top of the steel tube lining. An internal drop hammer can be used to tamp and consolidate the concrete but usually compressed air is the method employed — see Fig. II.14.

If waterlogged soil is encountered a pressure pile is usually formed by fixing to the head of the steel liner an air lock hopper through which the concrete can be introduced and consolidated whilst the bore hole remains under pressure in excess of the hydrostatic pressure of the ground water — see Fig. II.14.

Rotary bored piles: these can range from the short bored pile used in domestic dwellings (see Volume 1, Chapter 3) to the very large diameter piles used for concentrated loads in multi-storey buildings and bridge construction. The rotary bored pile is suitable for most cohesive soils such as clay and is formed using an auger which may be operated in conjunction with the steel tube liner according to the subsoil conditions encountered.

Two common augers are in use; the Cheshire auger which has 1½ to 2 helix turns at the cutting end and is usually mounted on a lorry or tractor. The turning shaft or kelly bar is generally up to 7.500 m long and is either telescopic or extendable. The soil is cut by the auger, raised to the surface and spun off the helix to the side of the bore hole from where it is removed from site — see Fig. II.15. Alternatively a continuous or flight auger can be used where the spiral motion brings the spoil to the surface for removal from site. Flight augers are usually mounted on an adapted excavator or crane power unit. Bucket boring tools, although not so common as augers, can also be used.

Large diameter bored piles are usually considered to be those over 750 mm and can be formed with diameters up to 2.600 m with lengths ranging from 24.000 to 60.000 m to carry loadings from 2 500 to 8 000 kN. They are suitable for use in stiff clays for structures having high concentrated loadings and can be lined or partially lined with steel tubes as required. The base or toe of the pile can be enlarged or underreamed up to three times the shaft diameter to increase the bearing capacity of the pile.

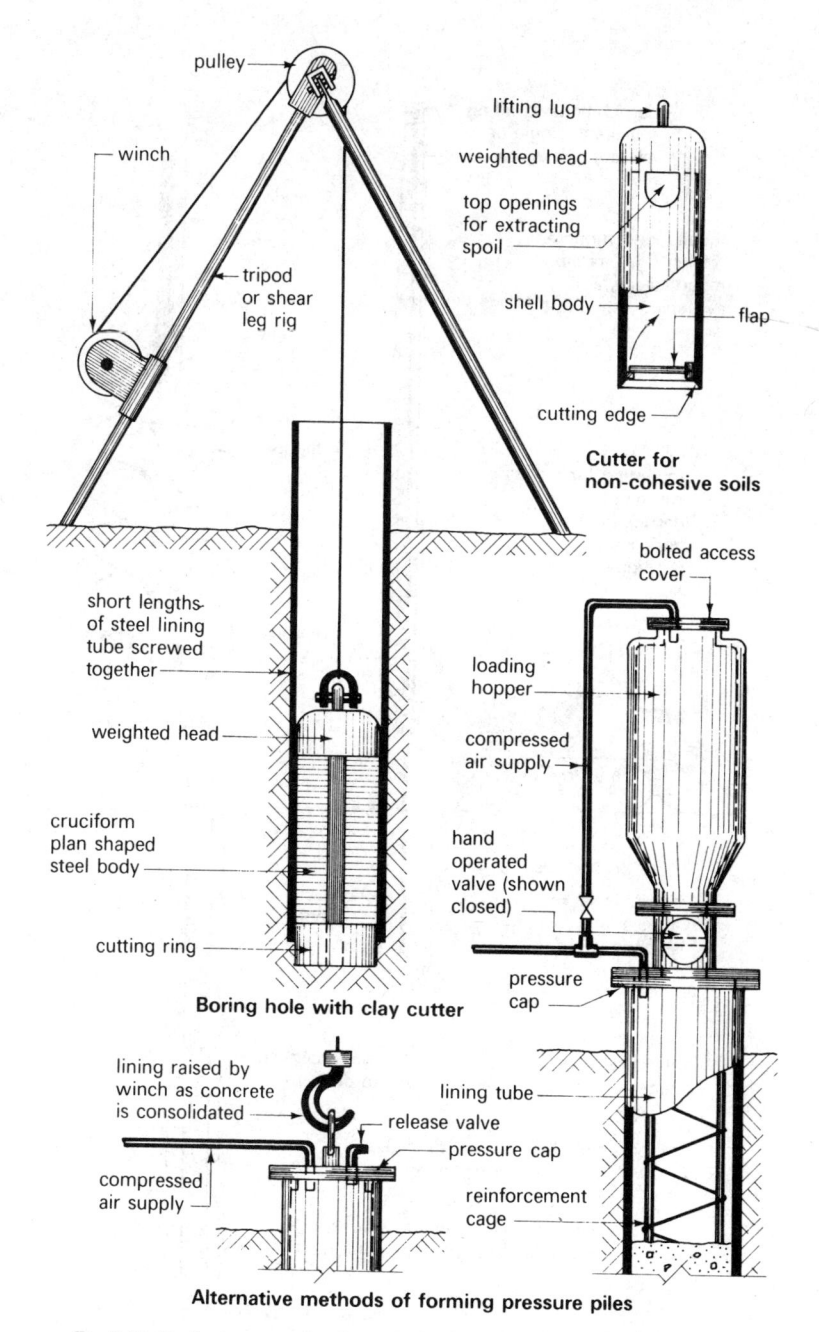

Fig II.14 Typical percussion bored pile details

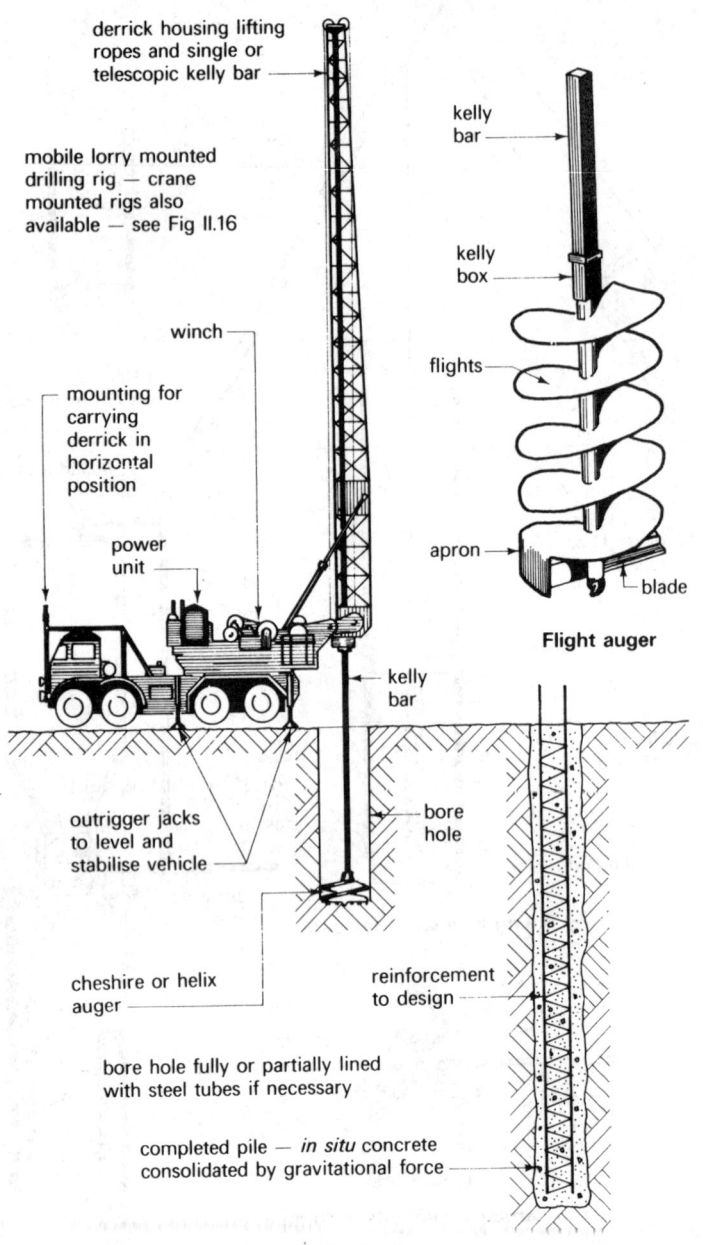

Fig II.15 Typical rotary bored pile details

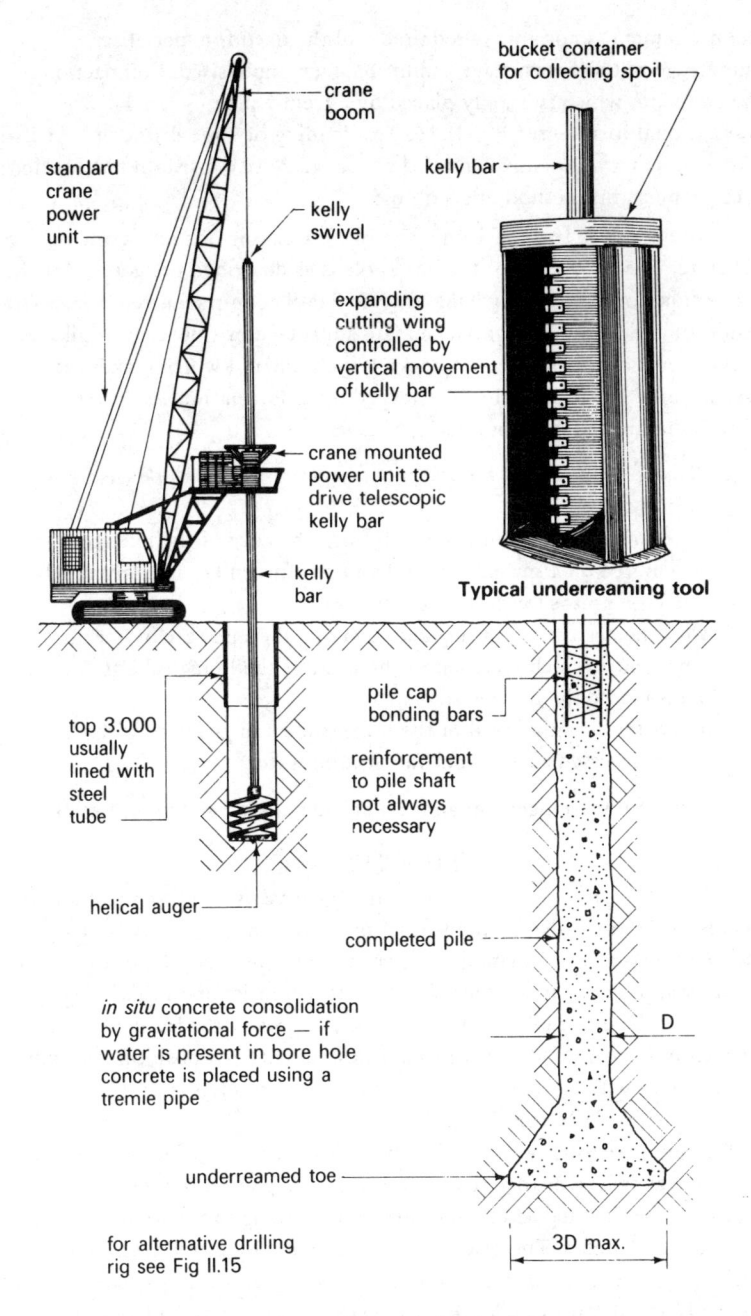

Fig II.16 Typical large diameter bored pile details

Reinforcement is not always required and the need for specialists' knowledge at the design stage cannot be over-emphasised. Compaction of the concrete, which is usually placed by a tremie pipe, is generally by gravitational force — see Fig. II.16. Test loading of large diameter bored piles can be very expensive and if the local authority insist on test loading it can render this method uneconomic.

Prestcore piles: a form of composite pile consisting of precast and *in situ* concrete. The formation of the bore hole is as described previously for the percussion bored pile using light, easy to handle equipment requiring a low headroom or working height. The main advantage of this form of pile lies in the fact that the problem of necking is eliminated which makes the system suitable for piling in waterlogged soils. Formation of a prestcore pile can be divided into four distinct stages:

1. *Boring* — lined bore hole formed by percussion methods using a tripod rig.
2. *Assembly* — precast units which form the core of the pile are assembled on a special mandrel and reinforcement is inserted before the core unit is lowered into position.
3. *Pressing the core* — raising and lowering the pile core by means of a pneumatic winch attached to the head of the lining tube to consolidate the bearing stratum.
4. *Grouting* — withdrawal of the lining tube and grouting with the aid of compressed air to expel any ground water.

The assembly arrangement and the unit details are shown in Fig. II.17.

PILE TESTING

The main objective of forming a test pile is to confirm that the design and formation of the chosen pile type is adequate. It is always advisable to form at least one test pile on any piling contract and indeed many local authorities will insist upon this being carried out. The test pile must not be used as part of the finished foundations but should be formed and tested in such a position that will not interfere with the actual contract but is nevertheless truly representative of site conditions.

Test piles are usually overloaded by at least 50% of the design working load to near failure or to actual failure depending upon the test data required. Any loading less than total failure loading should remain in place for at least 24 hours. The test pile is bored to the required depth or driven to the required set (which is a predetermined penetration per blow or series of blows) after which it can be tested by one of the following methods:

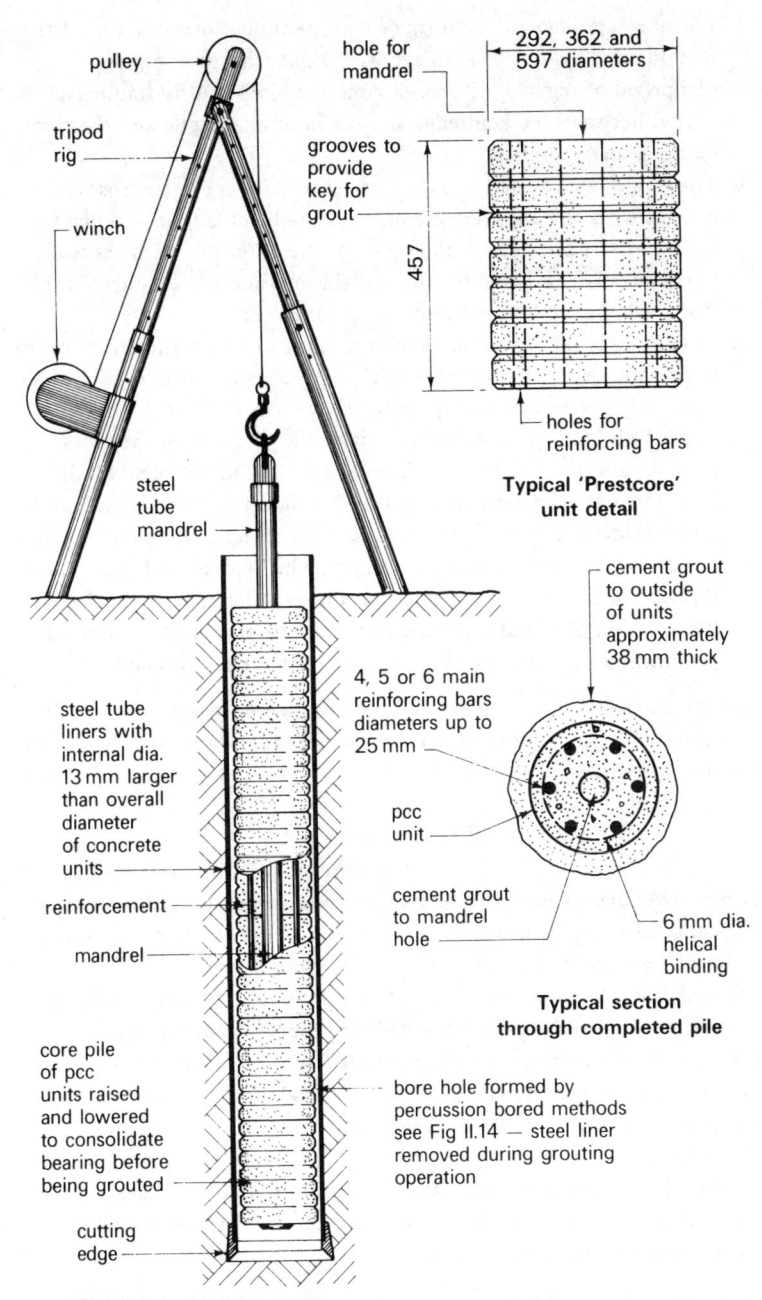

Fig II.17 BSP prestcore bored pile

1. Forming a grillage or platform of steel or timber over the top of the test pile and loading the grillage with a quantity of kentledge composed of pig iron or precast concrete blocks. A hydraulic jack is placed between the kentledge and the head of the pile and the test load gradually applied.
2. Three piles are formed and the outer two piles are tied across their heads with a steel or concrete beam. The object is to jack down the centre or test pile against the uplift of the outer piles. Unless two outer piles are available for the test this method can be uneconomic if special outer piles have to be bored or driven.
3. Anchor stressing wires are secured into rock or by some other means to provide the anchorage and uplift resistance to a cross beam of steel or concrete by passing the wires over the ends of the beam. The test load is applied by a hydraulic jack placed between the cross member and the pile head as described for the previous method.
4. Constant rate of penetration test is a method whereby the pile under test is made to penetrate the ground at a constant rate, generally in the order of about 0.8 mm per minute. The load needed to produce this rate of penetration is plotted against a deflection or timebase. When no further load is required to maintain this constant rate of penetration the ultimate bearing capacity has been reached.

It must be noted that the safe load for a single pile cannot necessarily be multiplied by the number of piles in a cluster to obtain the safe load of a group of piles.

PILE CAPS

Apart from the simple situations such as domestic dwellings or where large diameter piles are employed piles are not usually used singly but are formed into a group or cluster. The load is distributed over the heads of the piles in the group by means of a reinforced cast *in situ* concrete pile cap. To provide structural continuity the reinforcement in the piles is bonded into the pile cap; this may necessitate the breaking out of the concrete from the heads of the piles to expose the reinforcement. The heads of piles also penetrate the base of the pile cap some 100 to 150 mm to ensure continuity of the members.

Piles should be spaced at such a distance so that the group is economically formed and at the same time prevent an interaction between adjacent piles. Actual spacings must be selected upon subsoil conditions but the usual minimum spacings are:

1. *Friction piles* – 3 pile diameters or 1.000 m, whichever is greater.
2. *End bearing piles* – 2 pile diameters or 750 mm, whichever is greater.

Typical pile cap plans

Elevation showing typical pile cap and ground beam details

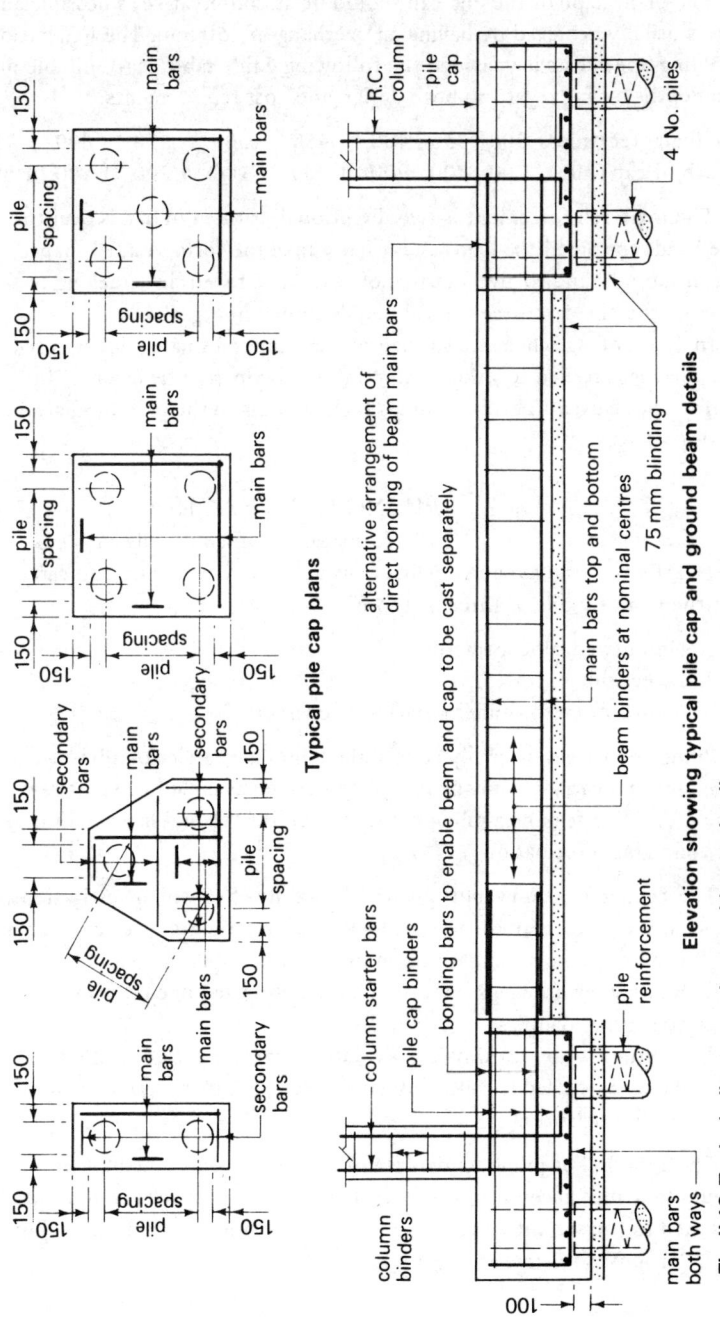

Fig II.18 Typical pile cap and ground beam details

The plan shape of the pile cap should be as conservative as possible and this is usually achieved by having an overhang of 150 mm. The Federation of Piling Specialists have issued the following guide table as to suitable pile cap depths having regard to both design and cost requirements:

Pile diameter (mm)	300	350	400	450	500	550	600	750
Depth of cap (mm)	700	800	900	1 000	1 100	1 200	1 400	1 800

The main reinforcement is two-directional, formed in bands over the pile heads to spread the loads and usually take the form of a 'U' shaped bar suitably bound to give a degree of resistance to surface cracking of the faces of the pile cap — see typical details shown in Fig. II.18.

In many piling schemes, especially where capped single piles are used, the pile caps are tied together with reinforced concrete tie beams. The beams can be used to carry loadings such as walls to the pile foundations — see Fig. II.18.

PILING CONTRACTS

The formation of piled foundations is a specialist's task and as such a piling contractor may be engaged to carry out the work by one of three methods:

1. Nominated subcontractor.
2. Direct subcontractor.
3. Main contractor under a separate contract.

Piling companies normally supply the complete service of advising, designing and carrying out the complete site works including setting out if required. When seeking a piling tender the builder should have and supply the following information:

1. Site investigation report giving full details of subsoil investigations, adjacent structures, topography of site and any restrictions regarding headroom, noise and vibration limitations.
2. Site layout drawings indicating levels, proposed pile positions and structural loadings.
3. Contract data regarding tender dates, contract period, completion dates, a detailed specification and details of any special or unusual contract clauses.

The appraisal of piling tenders is not an easy task and the builder should take into account not only costs but also any special conditions attached to the submitted tender and the acceptance of the local authority to the proposed scheme and system.

Part III
Frameworks

6
Portal frame theory

A portal frame may be defined as a continuous or rigid frame which has the basic characteristic of a rigid or restrained joint between the supporting member or column and the spanning member or beam. The object of this continuity of the portal frame is to reduce the bending moment in the spanning member by allowing the frame to act as one structural entity, thus distributing the stresses throughout the frame.

If a conventional simply supported beam was used (over a large span) an excessive bending moment would occur at mid-span which would necessitate a deep heavy beam or a beam shaped to give a large cross section at mid-span. Alternatively a deep cross member of lattice struts and ties could be used. The main advantage of the simply supported frame lies in the fact that the column loading is for all practicable purposes axial and therefore no bending is induced into the supporting members, which may well ease design problems since it would be statically determinate, but does not necessarily produce an economic structure. Furthermore the use of a portal frame eliminates the need for a lattice of struts and ties within the roof space, giving a greater usable volume to the structure and generally a more pleasing internal appearance.

The transfer of stresses from the beam to the column in rigid frames will require special care in the design of the joint between the members, similarly the horizontal thrust and/or rotational movement at the foundation connection needs careful consideration. Methods used to overcome excessive forces at the foundation are:

1. Reliance on the passive pressure of the soil surrounding the foundation.
2. Inclined foundations so that the curve of pressure is normal to the upper surface, thus tending to induce only compressive forces.
3. A tie bar or beam between opposite foundations.
4. Introducing a hinge or pin joint where the column connects to the foundation.

HINGES

Portal frames of moderate height and span are usually connected directly to their foundation bases forming rigid or unrestrained joints. The rotational movement caused by wind pressures tending to move the frames and horizontal thrusts of the frame loadings are generally resisted by the size of the base and the passive earth pressures. When the frames start to exceed 4.000 m in height and 15.000 m in span the introduction of a hinged or pin joint at the base connection should be considered.

A hinge is a device which will allow free rotation to take place at the point of fixity but at the same time will transmit both load and shear from one member to another. They are sometimes called pin joints, unrestrained joints and non-rigid joints. Since no bending moment is transmitted through a hinged joint the design is simplified by the structural connection becoming statically determinate. In practice it is not always necessary to provide a true 'pivot' where a hinge is included but to provide just enough movement to ensure the rigidity at the connection is low enough to overcome the tendency of rotational movement.

Hinges can be introduced into a portal frame design at the base connections and at the centre or apex of the spanning member, giving three basic forms of portal frame:

1. *Fixed or rigid portal frame* — all connections between frame members are rigid. This will give bending moments of lower magnitude more evenly distributed than other forms. Used for small to medium size frames where the moments transferred to the foundations will not be excessive.
2. *Two pin portal frame* — hinges are used at the base connections to eliminate the tendency of the base to rotate. The bending moments resisted by the supporting members will be greater than those encountered in the rigid portal frame. Main use is where high base moments and weak ground conditions are encountered.
3. *Three pin portal frame* — this form of frame has hinged joints at the base connections and at the centre of the spanning member. The

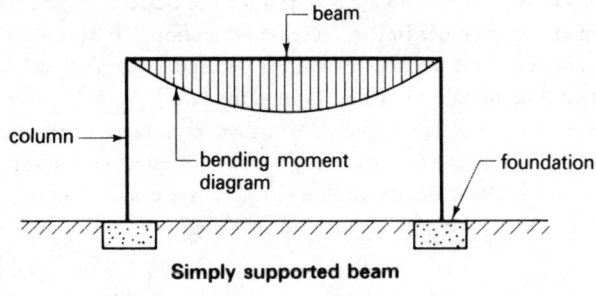

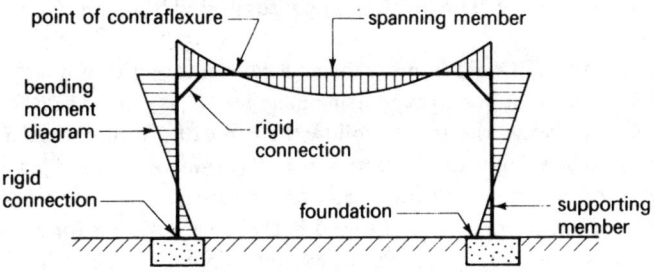

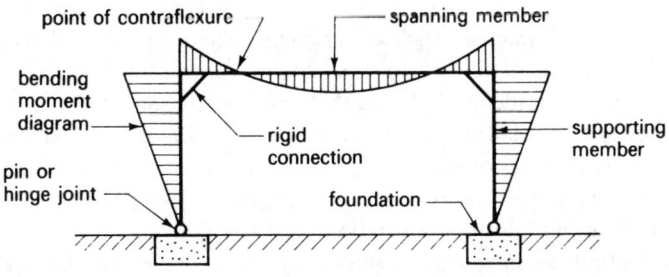

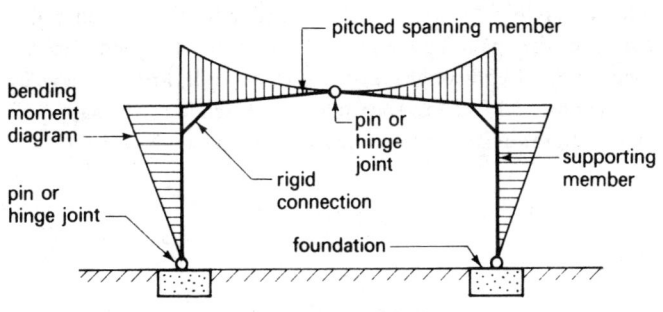

Fig III.1 Portal frames - comparison of bending moments

effect of the third hinge is to reduce the bending moments in the spanning member but to increase deflection. To overcome this latter disadvantage a deeper beam must be used or alternatively the spanning member must be given a moderate pitch to raise the apex well above the eaves level. Two other advantages of the three pin portal frame are that the design is simplified since the frame is statically determinate and on site they are easier to erect, particularly when preformed in sections.

A comparison of the bending moment diagrams for roof loads of the three forms of portal frame with a simply supported beam are shown in Fig. III.1.

Another form of rigid frame is the arch rib frame, which is not strictly a portal frame since it has no supporting members. The main design objective is to design the arch to follow the curve of pressure, thus creating a state of no bending when subjected to a uniformly distributed load. Any moments encountered with this form of frame are generally those induced by wind pressures. Hinges may be used in the same positions for the same reasons as described above for the conventional portal frames. The arch rib rigid frame is very often used where laminated timber is the structural material.

Most portal frames are made under factory controlled conditions off site which gives good dimensional and quality control but can create transportation problems. To lessen this problem and that of site erection splices may be used. These can be positioned at the points of contraflexure (see Fig. III.1), junction between spanning and supporting members and at the crown or apex of the beam. Most hinges or pin joints provide a point at which the continuity of fabrication is broken.

Portal frames constructed of steel, concrete or timber can take the form of the usual roof profiles used for single or multi-span buildings such as flat, pitched, northlight, monitor and arch. The frames are generally connected over the spanning members with purlins designed to carry and accept the fixing of lightweight roof coverings or deckings. The walls can be of similar material fixed to sheeting rails attached to the supporting members or alternatively clad with brick or infill panels.

7
Concrete portal frames

Concrete portal frames are invariably manufactured from high quality precast concrete suitably reinforced. In common with all precast concrete components for buildings, rapid advances in design and use were made after the Second World War due mainly to the shortage of steel and timber which prevailed at that time. In the main the use of precast concrete portal frames is confined to low pitch ($4°$ to $22½°$) single span frames but two storey and multi-span frames are available, giving a wide range of designs from only a few standard components.

The frames are generally designed to carry a lightweight (34 kg/m^2 maximum) roof sheeting or decking fixed to precast concrete purlins. Most designs have an allowance for snow loading of up to 73 kg/m^2 in addition to that allowed for the dead load of the roof covering. Wall finishes can be varied and intermixed since they are non-load bearing and therefore have to provide only the degree of resistance required for fire, thermal and sound insulation, act as a barrier to the elements and resist positive and negative wind pressures. Sheet claddings are fixed in the traditional manner, using hook bolts and purlins; sheet wall claddings are fixed in a similar manner to sheeting rails of precast concrete or steel spanning between or over the supporting members. Brick or block wall panels either of solid or cavity construction can be built off a ground beam constructed between the foundation bases or alternatively they can be built off the ground floor slab. It must be remembered that all such claddings must comply with any relevant Building Regulations.

FOUNDATIONS AND FIXINGS

The foundations for a precast concrete portal frame usually consist of a reinforced concrete isolated base or pad designed to suit loading and ground bearing conditions. The frame can be connected to the foundations by a variety of methods:

1. *Pocket connection* — the foot of the supporting member is located and housed in a void or pocket formed in the base so that there is an all round clearance of 25 mm to allow for plumbing and final adjustment before the column is grouted into the foundation base.
2. *Base plate connection* — a steel base plate is welded to the main reinforcement of the supporting member, or alternatively it could be cast into the column using fixing lugs welded to the back of the base plate. Holding down bolts are cast into the foundation base; the erection and fixing procedure follows that described for structural steelwork (see Volume 2, part III).
3. *Pin joint or hinge connection* — a special base or bearing plate is bolted to the foundation and the mechanical connection is made when the frames are erected — see Fig. III.5.

The choice of connection method depends largely upon the degree of fixity required and the method adopted by the manufacturer for his particular system.

ADVANTAGES

The main advantages of using precast concrete portal frames can be enumerated thus:

1. Factory production will result in accurate and predictable components since the criteria for design, quality and workmanship recommended in BS 8110 can be more accurately controlled under factory conditions than casting components *in situ*.
2. Most manufacturers produce a standard range of interchangeable components which, within the limitations of their systems, gives a well-balanced and flexible design range covering most roof profiles, single span frames, multi-span frames and lean-to roof attachments. By adopting this limited range of members the producers of precast portal frames can offer their products at competitive rates coupled with reasonable delivery periods.
3. Maintenance of precast concrete frames is not usually required unless the building owner chooses to paint or clad the frames.
4. Precast concrete products have their own built-in natural resistance to fire and therefore no fire-resistant treatment is required. By

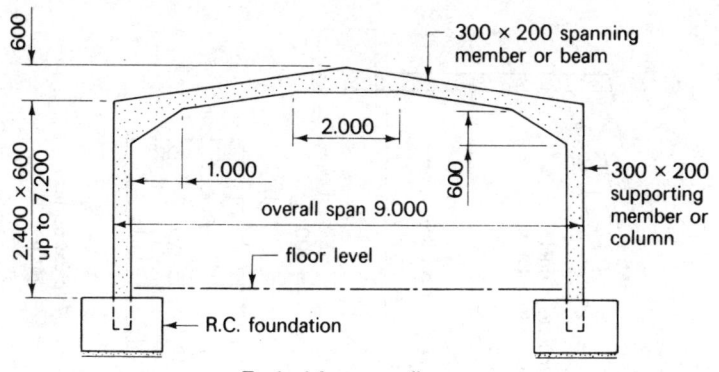

Typical frame outline

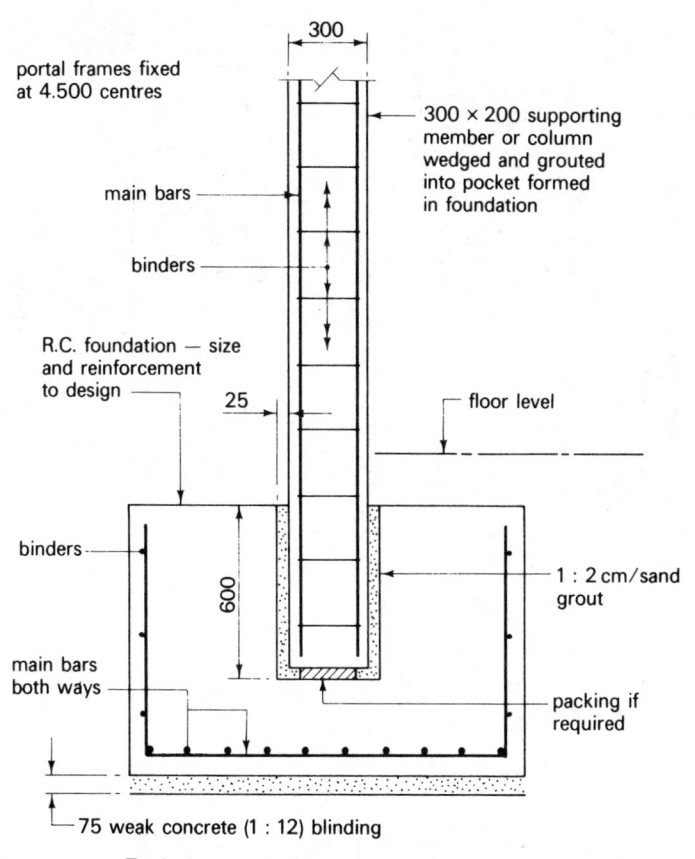

- 300 × 200 spanning member or beam
- 2.000
- 1.000
- overall span 9.000
- 600
- 300 × 200 supporting member or column
- 2.400 × 600 up to 7.200
- 600
- floor level
- R.C. foundation

Typical column to foundation connection

- portal frames fixed at 4.500 centres
- 300
- main bars
- 300 × 200 supporting member or column wedged and grouted into pocket formed in foundation
- binders
- R.C. foundation — size and reinforcement to design
- 25
- floor level
- binders
- 600
- 1 : 2 cm/sand grout
- main bars both ways
- packing if required
- 75 weak concrete (1 : 12) blinding

Fig III.2 Typical single span pcc frame

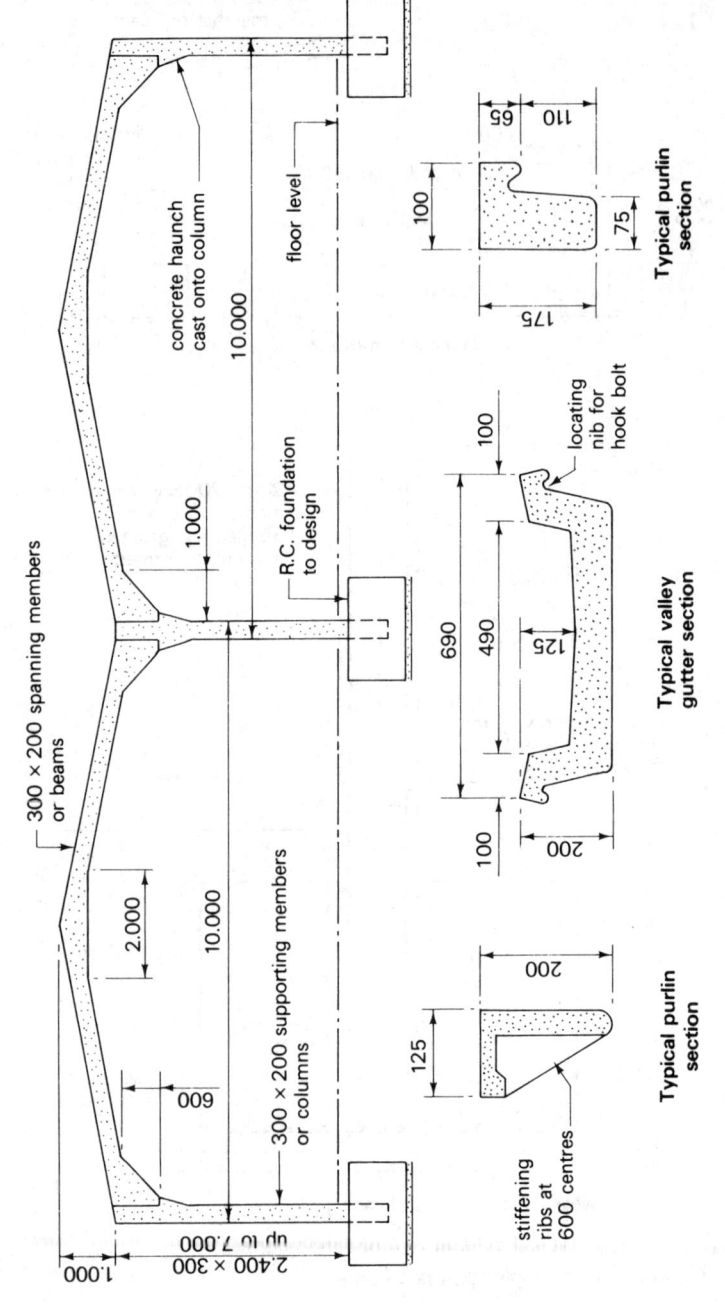

Fig III.3 Typical multi-span precast concrete portal frame

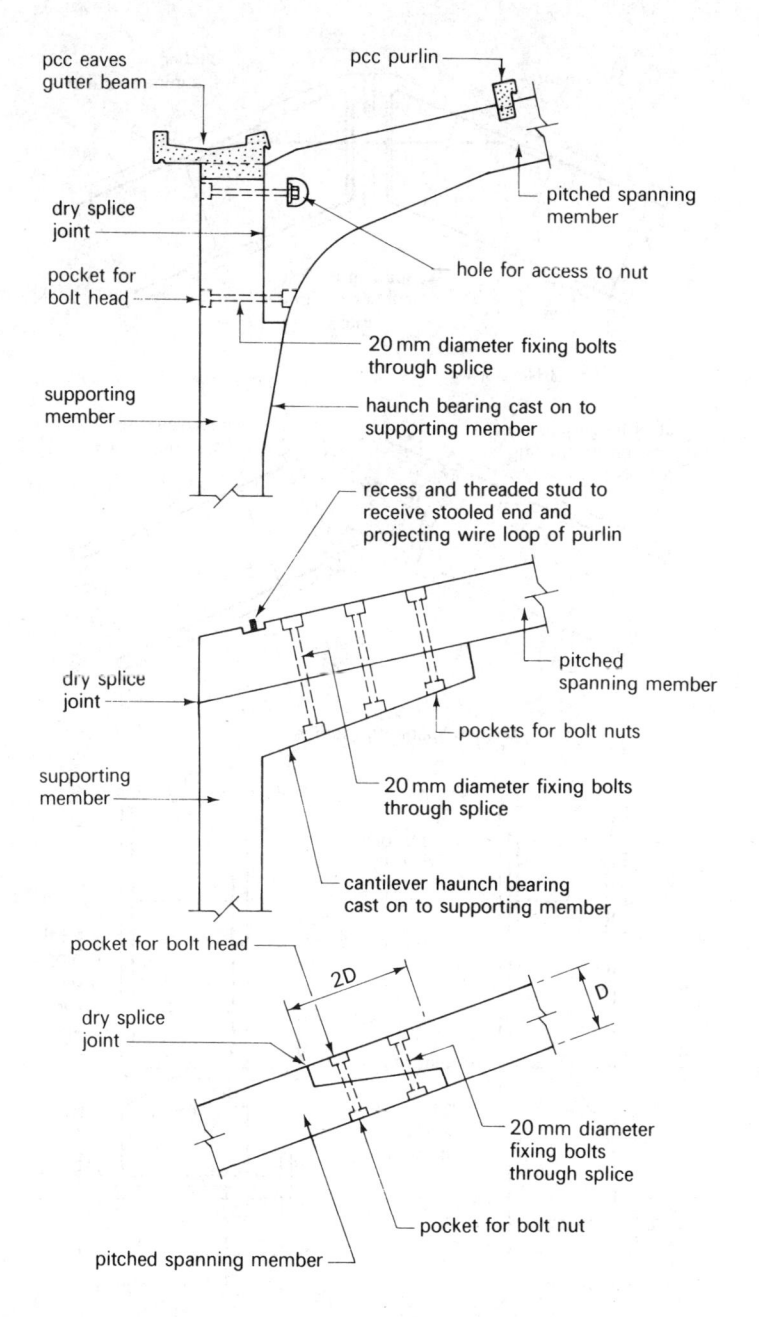

Fig III.4 Typical splice details for pcc portal frames

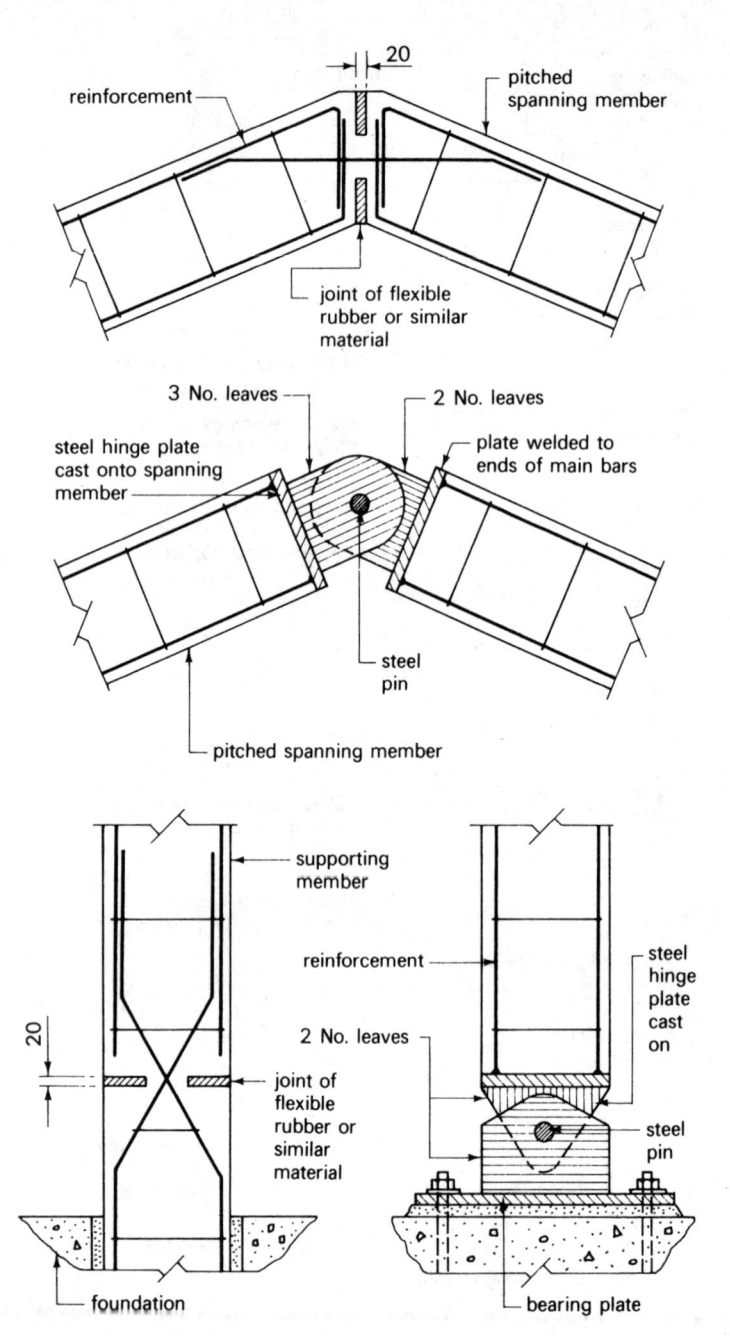

Fig III.5 Typical hinge details for pcc portal frames

varying the cover of concrete over the reinforcement most frames up to 24.000 m span are given a 1-hour fire resistance and frames exceeding this span are rated at 2-hour fire resistance.
5. The wind resistance of precast concrete portal frames to both positive and negative pressures is such that wind bracing is not usually required.
6. Where members of the frame are joined or spliced together the connections are generally mechanical (nut and bolt) and therefore the erection and jointing can be carried out by quickly trained semi-skilled labour.
7. The clean lines of precast concrete portal frames are considered to be aesthetically pleasing.
8. In most cases the foundation design, setting out and construction can be carried out by the portal frame sub-contracting firm.

Typical details of single span frames, multi-span frames, cladding supports, splicing and hinges are shown in Figs. III.2 to III.5.

8
Steel portal frames

Steel portal frames can be fabricated from standard universal beam, column and box sections. Alternatively a lattice construction of flats, angles or tubulars can be used. Most forms of roof profiles can be designed and constructed giving a competitive range when compared with other materials used in portal frame construction. The majority of systems employ welding techniques for the fabrication of components which are joined together on site using bolts or welding. An alternative system uses special knee joint, apex joint and base joint components which are joined on site to square cut standard beam or column sections supplied by the main contractor or by the manufacturer producing the jointing pieces.

The frames are designed to carry lightweight roof coverings of the same loading conditions as those given previously for precast concrete portal frames. Similarly wall claddings can be of the same specification as for precast concrete portal frames and fixed in the same manner. Any relevant Building Regulations must be observed and if the usage of the building, irrespective of the framing material, is for an industrial process the roof would have to comply with the requirements of The Building Regulations 1985 Schedule 1, Part L (see Part VI — roofs).

FOUNDATIONS AND FIXINGS

The foundation is usually a reinforced concrete isolated base or pad foundation designed to suit loading and ground bearing conditions. The connection of the frame to the foundation can be by one of three basic methods:

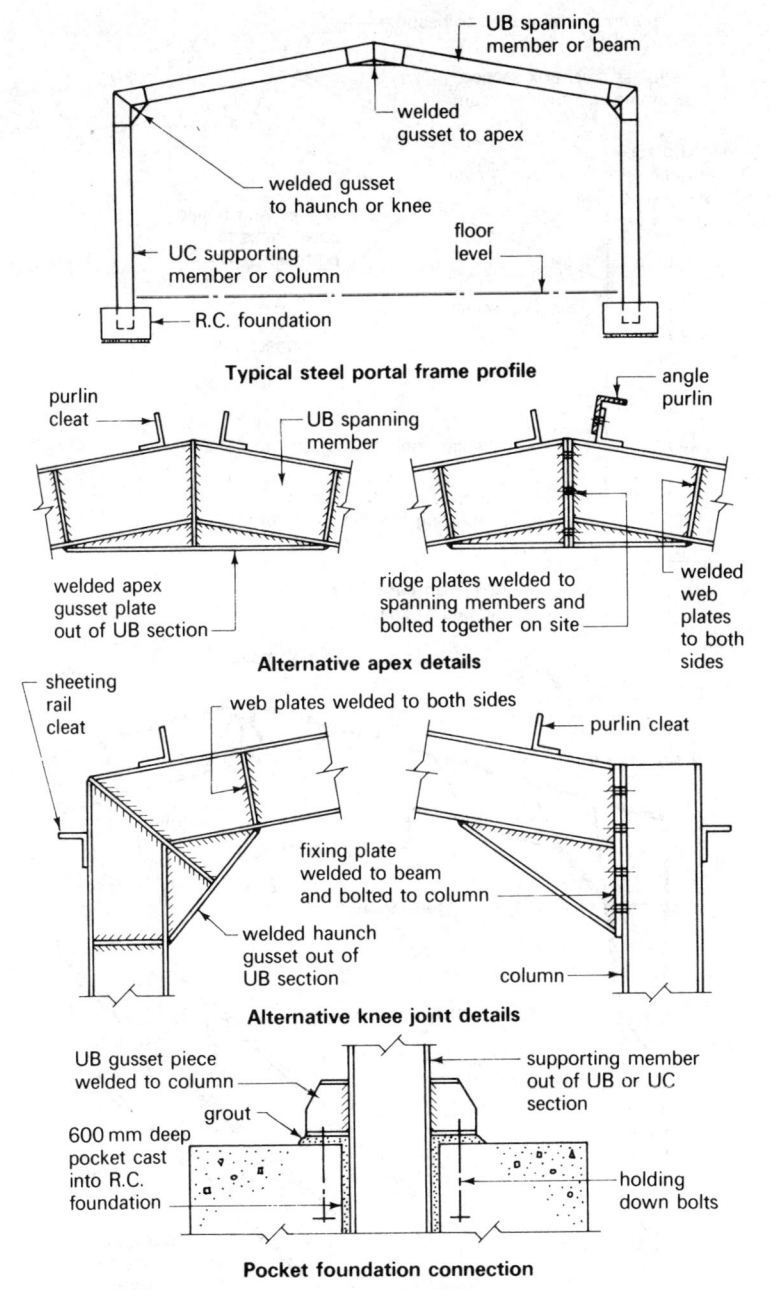

Fig III.6 Typical steel portal frame details

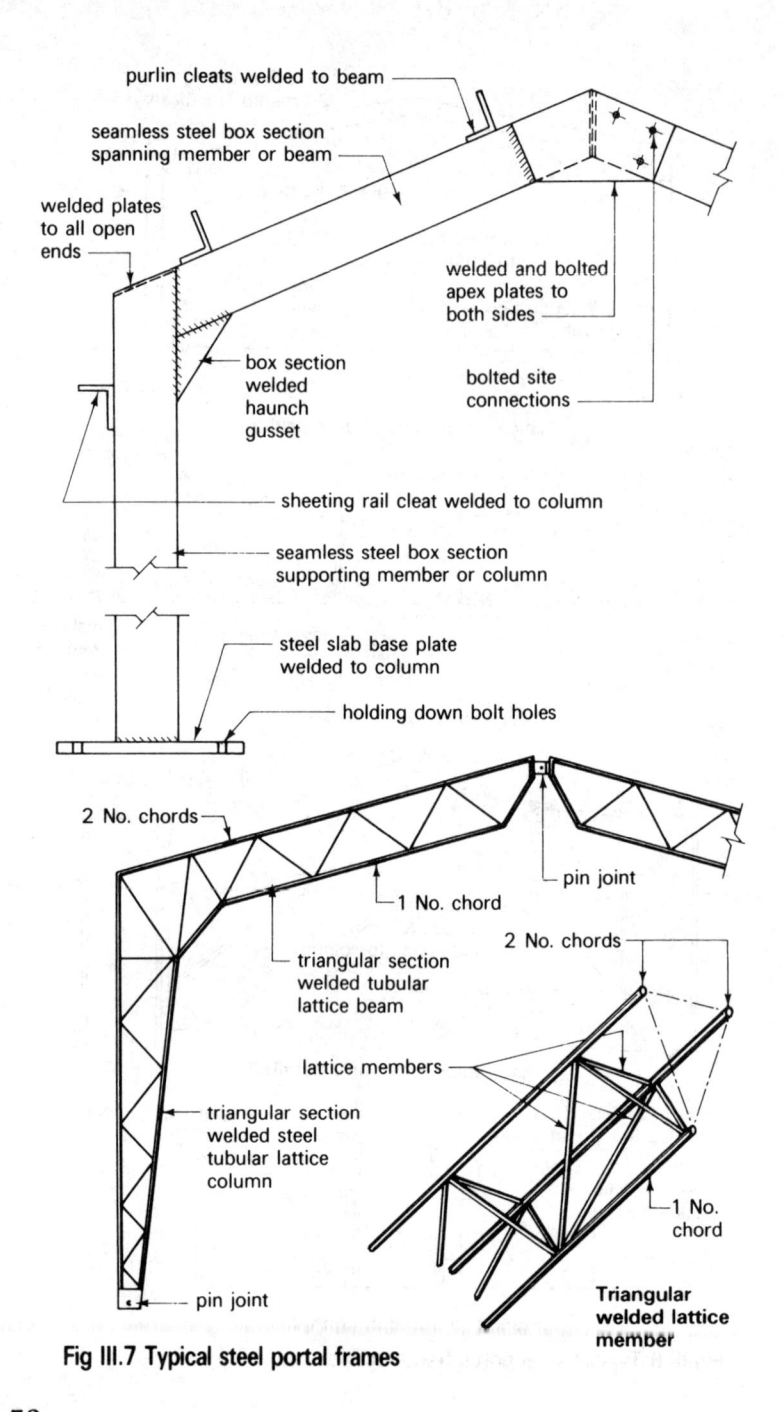

Fig III.7 Typical steel portal frames

Typical apex hinge joint

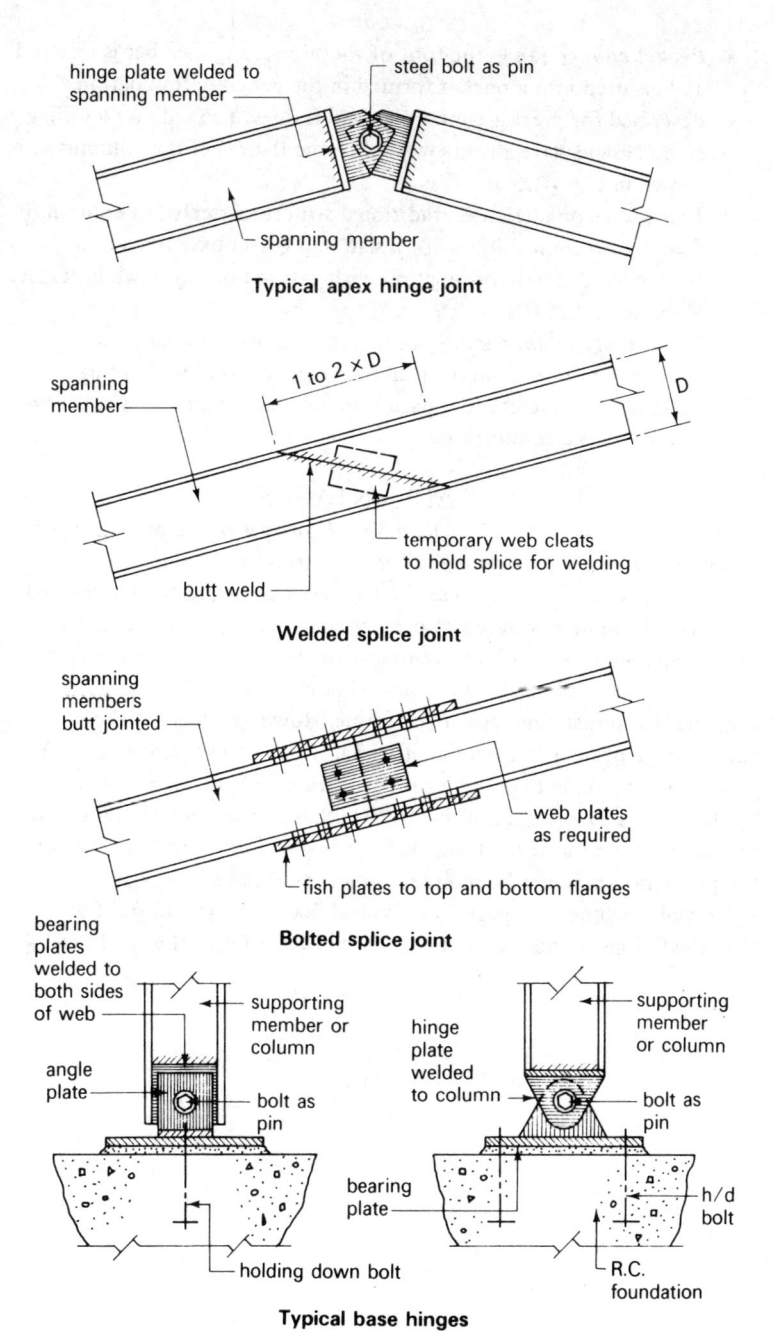

- hinge plate welded to spanning member
- steel bolt as pin
- spanning member

Welded splice joint

- spanning member
- 1 to 2 × D
- D
- temporary web cleats to hold splice for welding
- butt weld

Bolted splice joint

- spanning members butt jointed
- web plates as required
- fish plates to top and bottom flanges

Typical base hinges

- bearing plates welded to both sides of web
- angle plate
- supporting member or column
- bolt as pin
- holding down bolt
- hinge plate welded to column
- supporting member or column
- bolt as pin
- bearing plate
- h/d bolt
- R.C. foundation

Fig III.8 Steel portal frames - splices and hinges

1. *Pocket connection* — the foot of the supporting member is inserted and grouted into a pocket formed in the concrete foundation as described for precast concrete portal frames. To facilitate levelling some designs have gussets welded to the flanges of the columns as shown in Fig. III.6.
2. *Base plate connection* — traditional structural steelwork column to foundation connection using a slab or a gusset base fixed to a reinforced concrete foundation with cast in holding down bolts (see Volume 2, Part III).
3. *Pin or hinge connection* — special bearing plates designed to accommodate true pin or rocker devices are fixed by holding down bolts to the concrete foundation to give the required low degree of rigidity at the connection.

ADVANTAGES

The main advantages of factory controlled production are: a standard range of manufacturer's systems, a frame of good wind resistance and the ease of site assembly using quickly trained semi-skilled labour attributed to precast concrete portal frames can be equally applied to steel portal frames. A further advantage of steel is that generally the overall dead load of a steel portal frame is less than a comparable precast concrete portal frame. However, steel has the disadvantage of being a corrosive material which will require a long life protection of a patent coating or regular protective maintenance generally by the application of coats of paint. Steel has a lower fire resistance than precast concrete but if the frame is for a single storey building structural fire protection may not be required under the Building Regulations (see Approved Document B, page 37). Typical details of steel portal frames, cladding fixings, splicing and hinges are shown in Figs. III.6 to III.8.

9
Timber portal frames

Timber portal frames can be manufactured by several methods which produce a light, strong frame of pleasing appearance which renders them suitable for buildings such as churches, halls and gymnasiums where clear space and appearance are important. The common methods used are glued laminated portal frames, plywood faced portal frames and timber portal frames using solid members connected together with plywood gussets.

GLUED LAMINATED PORTAL FRAMES

The main objective of forming a laminated member consisting of glued layers of thin section timber members is to obtain an overall increase in strength of the complete component over that which could be expected from a similar sized solid section of a particular species of timber. This type of portal frame is usually manufactured by a specialist firm since the jigs required would be too costly for small outputs. The selection of suitable quality softwoods of the right moisture content is also important for a successful design. In common with other timber portal frames, these can be fully rigid, 2 pin or 3 pin structures.

Site work is simple, consisting of connecting the foot of the supporting member to the metal shoe fixing or to a pivot housing bolted to the concrete foundation and connecting the joint at the apex or crown with a bolt fixing or a hinge device. Most glued laminated timber portal frames are fabricated in two halves which eases transportation problems and gives maximum usage of the assembly jigs. The frames can be linked together at

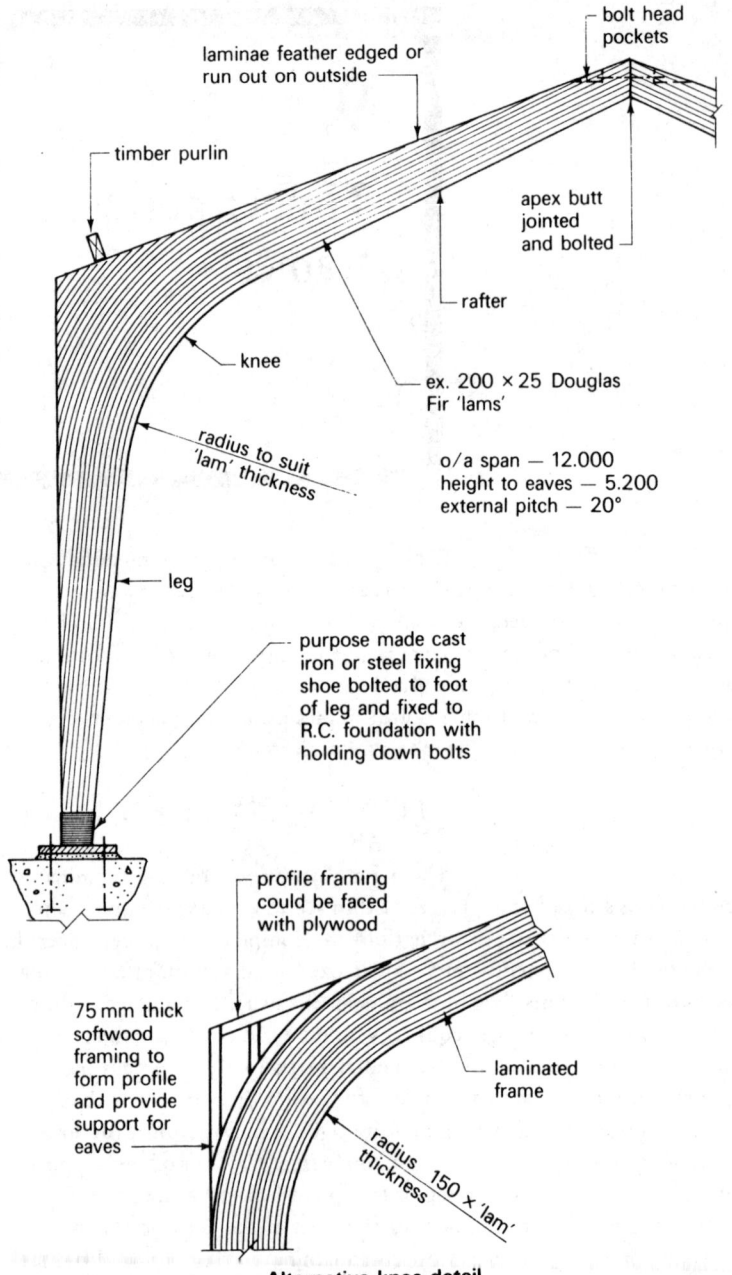

- bolt head pockets
- laminae feather edged or run out on outside
- timber purlin
- apex butt jointed and bolted
- rafter
- knee
- ex. 200 × 25 Douglas Fir 'lams'
- radius to suit 'lam' thickness

o/a span — 12.000
height to eaves — 5.200
external pitch — 20°

- leg
- purpose made cast iron or steel fixing shoe bolted to foot of leg and fixed to R.C. foundation with holding down bolts
- profile framing could be faced with plywood
- 75 mm thick softwood framing to form profile and provide support for eaves
- laminated frame
- radius 150 × 'lam' thickness

Alternative knee detail

Fig III.9 Typical glued laminated portal frame

roof level with timber purlins and clad with a lightweight sheeting or decking; alternatively, they may be finished with traditional roof coverings. Any form of walling can be used in conjunction with these frames provided such walling forms comply with any of the applicable Building Regulations. Typical details are shown in Fig. III.9.

PLYWOOD FACED PORTAL FRAMES

These frames are suitable for small halls, churches and schools with spans in the region of 9.000 m The portal frames are in essence boxed beams consisting of a skeleton core of softwood members faced on both sides with plywood which takes the bending stresses. The hollow form of the construction enables electrical and other small services to be accommodated within the frame members. Design concepts, fixing and finishes are as given above for glued laminated portal frames. Typical details are shown in Fig. III.10.

SOLID TIMBER AND PLYWOOD GUSSETS

These frames were developed to provide a simple and economic timber portal frame for clear span buildings using ordinary tools and basic skills. The general concept of this form of frame varies from the two types of timber portal frames previously described in that no glueing is used, the frames are spaced close together (600, 900 and 1 200 mm centres) and are clad with a plywood sheath so that the finished structure acts as a shell giving a lightweight building which is very rigid and strong. The frames can be supplied in two halves and assembled by fixing the plywood apex gussets on site before erection or alternatively they can be supplied as a complete frame ready for site erection.

The foundations for this form of timber portal frame consists of a ground beam or alternatively the frames can be fixed to the edge of a raft slab. A timber spreader or sole plate is used along the entire length of the building to receive and distribute the thrust loads of the frames. Connection to this spreader plate is made by using standard galvanised steel joists hangers or by using galvanised steel angle cleats. Standard timber windows and doors can be inserted into the side walls by trimming in the conventional way and infilling where necessary with studs, noggins and rafters. Typical details are shown in Fig. III.11

The advantages of all timber portal frame types can be enumerated as follows:

1. Constructed from readily available materials at an economic cost.
2. Light in weight.

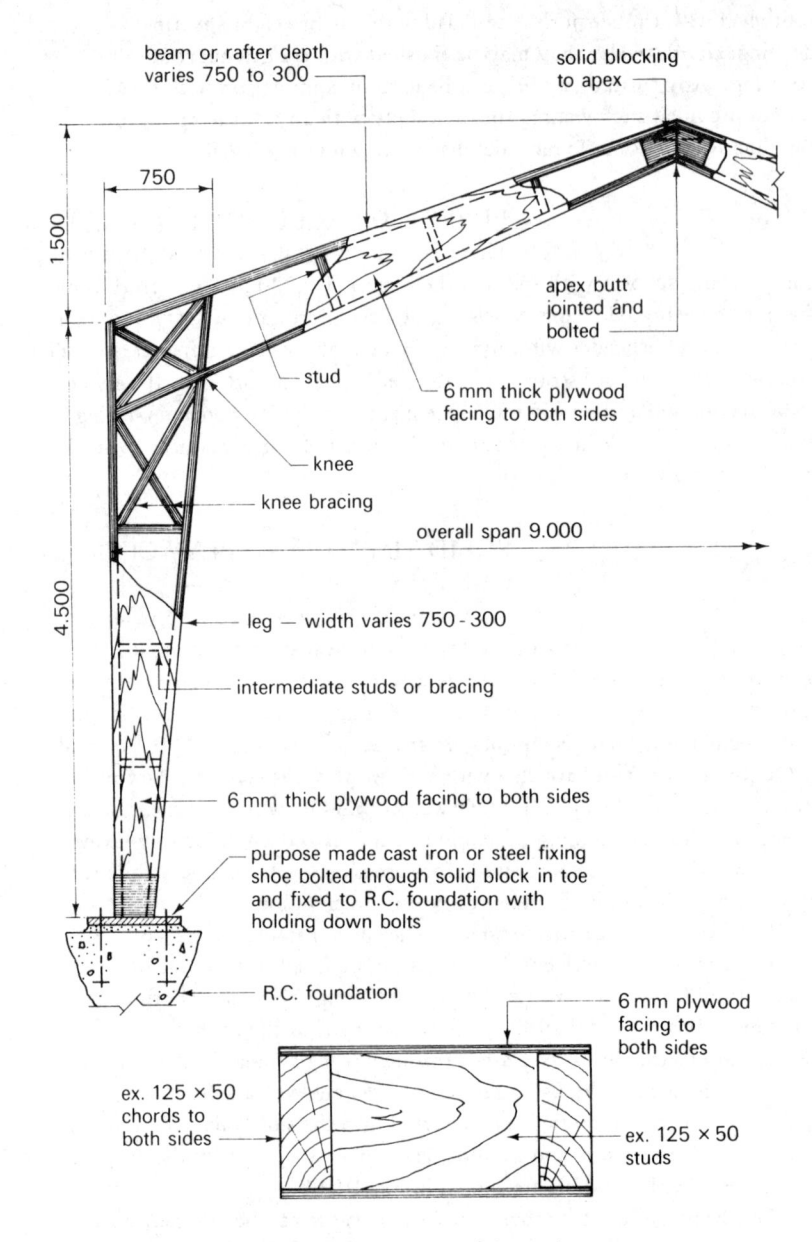

Fig III.10 Typical plywood faced portal frame

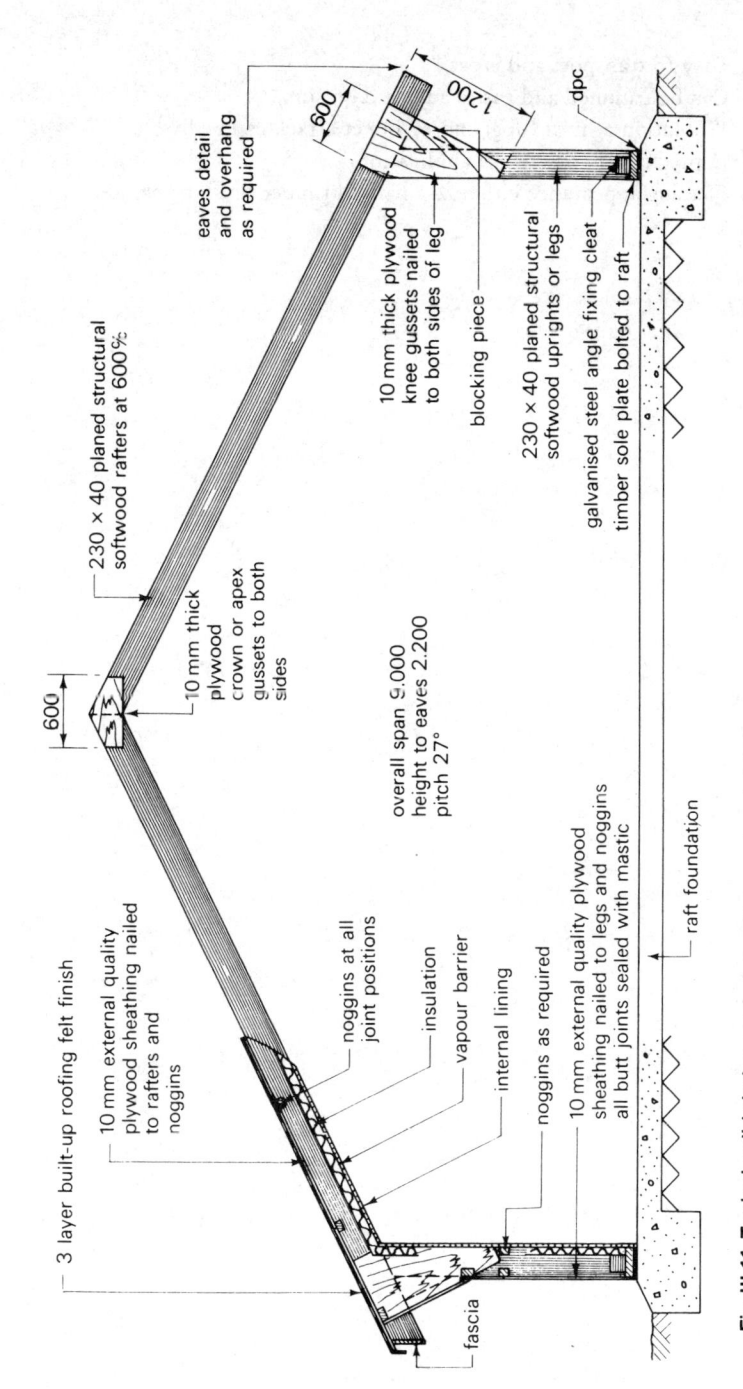

Fig III.11 Typical solid timber and plywood gusset portal frame

3. Easy to transport and erect.
4. Can be trimmed and easily adjusted on site.
5. Protection against fungi and/or insect attack can be by impregnation, or surface application.
6. Pleasing appearance either as a natural timber finish or painted.

Part IV
Fire

10
The problem of fire

Since early times fire has been one of man's greatest aids to his advancement; it gives him a source of both heat and light. Today fire is still of great benefit to man's wellbeing if it is controlled, but if allowed to start and spread without strict control it can be one of the greatest hazards man has to face. Early man considered fire to be a natural element like air and water; later experimenters found that the residue of a burnt fuel (ash) weighed less than the fuel before it was burnt and concluded that some substance was removed during the combustion period; this they called phlogiston after the Greek word phlogistos, meaning inflammable. The doctrine of phlogistics was overthrown by a French chemist Antoine Lavoisier (1743-1794), who became known as the father of modern chemistry.

Lavoisier discovered by his researches and experiments that air consists of $\frac{1}{5}$th oxygen and that the other main gas, nitrogen, accounted for the bulk of the remaining $\frac{4}{5}$ths. He showed that oxygen played an important part in the process of combustion and that nitrogen does not support combustion. This discovery of the true nature of fire led to the conclusion that fire is a chemical reaction whereby atoms of oxygen combined with other atoms such as carbon and hydrogen, releasing water, carbon dioxide and energy in the form of heat. The chemical reaction will only start at a suitable temperature, which varies according to the substance or fuel involved. During combustion gases will be given off, some of which are more inflammable than the fuel itself and therefore ignite and appear as flames, giving light which is due to tiny particles being heated to a point at which they glow. Smoke is an indication of incomplete combustion and can give rise to deposits of solid carbon commonly known as soot.

From the discovery of the true nature of fire and processes of combustion it can be concluded that there are three essentials to all fires:

1. *Fuel* — generally any organic material is suitable.
2. *Heat* — correct temperature to promote combustion of a particular fuel. Heat can be generated deliberately, which is termed ignition, or it can be spontaneous when the fuel itself ignites.
3. *Oxygen* — air is necessary to sustain and support the combustion process.

The above is often referred to as the triangle of fire: remove any one of the three essentials and combustion cannot take place. This fact provides the whole basis for fire prevention, fire protection and fire fighting. If non-combustible materials were used in the construction and furnishing of buildings fires would not develop. This method is far too restrictive on the designer and builder therefore combustible materials are used and protected with layers or coverings of non-combustible materials where necessary. Fire fighters try to remove one side of the fire triangle; to remove the fuel is not generally practicable but by using a cooling agent such as water the heat can be reduced to a safe level, or, alternatively, by using a blanketing agent the supply of oxygen can be cut off and fire extinguished.

The cost to the nation resulting from fires in buildings is very difficult to assess but a figure in excess of £1 000 per minute for every day of the year amounting to an annual cost of over £500 000 000 is not unrealistic The cost to manufacturers and employers in terms of loss of goodwill, loss of production, effect of fatalities and injuries to employees and the delay in returning to full production or working are almost incalculable. Therefore it can be seen that the seriousness of fires within buildings cannot be overstated.

Fire has no respect for persons or places, it can and does break out in all forms of buildings. The common belief that 'it never happens to me, only to others' is one of the major factors. The disregard by people and firms of even elementary precautions to obvious fire hazards is clearly shown by the UK fire and loss statistics published annually by Her Majesty's Stationery Office. These statistics show that private dwellings are most vulnerable, accounting usually for approximately 50% of all fires. Another high risk area is the distributive trades where large quantities of inflammable goods are held in store. One of the main causes of fires is faulty electrical equipment and wiring. This is not an indication that the plant or installation is poor but that maintenance, renewal and routine checks carried out by experienced personnel is in many cases non-existent. In domestic dwellings the same argument can be applied to gas services and installations. Another and indeed frightening statistic is the high number of fires (in excess of 3 000) caused annually in dwellings by children playing with fire, matches and other easily ignitable home gadgets. This could be the result of lack of education as to the dangers or lack of parental control.

Obviously designers and builders alike cannot be held responsible for the actions or non-actions of the occupants of the buildings they create but they can ensure that these structures are designed and constructed in such a manner that they give the best possible resistance to the action of fire should it occur.

The precautions which can be taken within buildings to prevent a fire occurring, or if it should occur of containing it within the region of the outbreak, providing a means of escape for people in the immediate vicinity and fighting the fire can be studied under three headings:

1. Structural fire protection.
2. Means of escape in case of fire.
3. Fire fighting.

The latter, which is generally integrated with the services of a building, is usually considered in that context and therefore apart from passing references no deep study of this aspect is included in this text.

11
Structural fire protection

The purpose of structural fire protection is to ensure that during a fire the temperature of structural members or elements does not increase to a figure at which their strength would be adversely affected. It is not practicable or possible to give an element complete protection in terms of time, therefore elements are given a fire resistance for a certain period of time which it is anticipated will give sufficient delay to the spread of fire, ultimate collapse of the structure, time for persons in danger to escape and to enable fire fighting to be commenced.

Before a fire-resistance period can be determined it is necessary to consider certain factors:

1. Fire load of the building.
2. Behaviour of materials under fire conditions.
3. Behaviour of combinations of materials under fire conditions.
4. Building Regulation requirements as laid down in Part B.

FIRE LOAD

Buildings can be graded as to the amount of overall fire resistance required by taking into account the following:

1. Size of building.
2. Use of building.
3. Fire load.

The fire load is an assessment of the severity of a fire due to the combustible materials within a building. This load is expressed as the amount of heat which would be generated per unit area by the complete combustion of its contents and combustible members and is given in Joules per square metre. It should be noted that the numerical grade is equivalent to the minimum number of hours fire resistance which should be given to the elements of the structure.

Grade 1 — Low fire load, not more than 1 150 MJ/m^2. Typical buildings within this grade are flats, offices, restaurants, hotels, hospitals, schools, museums and public libraries.

Grade 2 — Moderate fire load, 1 150 to 2 300 MJ/m^2. Typical examples are retail shops, factories and workshops.

Grade 4 — High fire load, 2 300 to 4 600 MJ/m^2. Typical examples are certain types of workshops and warehouses.

When deciding the grade no account is taken of the effects of any permanent fire protection installations such as sprinkler systems. The above principles are incorporated into the Building Regulations and in particular Part B.

FIRE RESISTANCE OF MATERIALS

The materials used in buildings can be studied as separate entities as to their behaviour when subjected to the intense heat encountered during a fire and as to their ability to spread fire over their surfaces.

Structural steel is not considered to behave well under fire conditions although its ability to spread fire over its surface is negligible. As the fire progresses and the temperature of steel increases there is an actual gain in the ultimate strength of mild steel. This gain in strength decreases back to normal over the temperature range of 250 to 400°C. The decrease in strength continues and by the time the steel temperature has reached 550°C it will have lost most of its useful strength. Since the rise in temperature during the initial stages of a fire is rapid this figure of 550°C can be reached very quickly. If the decrease in strength results in the collapse of a member the stresses it was designed to resist will be redistributed; this could cause other members to be overstressed and progressive collapse could occur.

Reinforced concrete structural members have good fire resistance properties, and being non-combustible do not contribute to the spread of flame over their surfaces. It is possible however under the intense and prolonged heat of a fire that the bond between the steel reinforcement and the concrete will be broken. This generally results in spalling of the concrete which decreases both the protective cover of the concrete over the steel and the cross sectional area. Like structural steel members, this can result in a redistribution of stresses leading to overloading of certain members, culminating in progressive collapse.

Timber, strange as it may seem, behaves very well structurally under the action of fire. This is due to its slow combustion rate, the strength of its core failure remaining fairly constant. The ignition temperature of timber is low (250-300°C) but during combustion the timber chars at an approximate rate of 0.5 mm per minute, the layer of charcoal so formed slows down the combustion rate of the core. Although its structural properties during a fire are good, timber being an organic material and therefore combustible, will spread fire over its surface which makes it unsuitable in most structural situations without some form of treatment.

From the above brief considerations it is obvious that designers and builders need to have data on the performance, under the conditions of fire, of materials and especially combinations of materials forming elements. Such information is available in BS 476 which is divided into a number of parts which relate to the various fire tests applied to building materials and structures.

FIRE TESTS — BS 476

BS 476 consists of nine parts numbered 3 to 8, 10, 11 and 31. Part 1 has been replaced by part 8 and Part 2 has been incorporated in BS 2782 — Methods of Testing Plastics. Parts 3 to 8 are very significant and worth noting.

Part 3 External Fire Exposure Roof Tests

A series of tests for grading roof structures in terms of time for:

1. Resistance to external penetration by fire.
2. Distance of spread of flame over the external surface under certain conditions.

The tests are applied to a specimen of roof structure 838 mm (33 inches) square which represents the actual roof construction including at least one specimen of any joints used and complete with any lining which is an integral part of the construction.

Three tests are applied:

1. Preliminary ignition test.
2. Fire penetration test.
3. Roof surface spread of flame test.

After testing the specimen or form of roof construction receives a lettered designation thus:

First letter (penetration)

A. No penetration within 1 hour.
B. Specimen penetrated in not less than 1½ hours.
C. Specimen penetrated in less than ½ hour.
D. Specimen penetrated in the preliminary flame test

Second letter (spread of flame)

A. No spread of flame.
B. Not more than 533 mm (21 inches) spread.
C. More than 533 mm spread.
D. Specimens which continue to burn for 5 minutes after withdrawal of the test flame or spread more than 381 mm (15 inches) across the region of burning in the preliminary flame test.

Specimens can be tested as an inclined or flat structure and are prefixed EXT.S or EXT.F accordingly. If during the test any dripping from the underside of the specimen, any mechanical failure and/or development of holes is observed a suffix 'X' is added to the designation thus:
EXT.S.AB–EXT.S.ABX

Part 4 Non-combustibility Test

Generally organic materials are combustible, whereas inorganic materials are non-combustible which can be defined as a material not capable of undergoing combustion. For the purposes of BS 476 materials used in the construction and finishing of buildings or structures are classified as 'non-combustible' or 'combustible' according to their behaviour in the non-combustibility test.

A material is classified by this test as non-combustible if none of the three specimens tested either:

1. Causes the temperature reading from either of the two thermocouples used to rise by $50°C$ or more above the initial furnace temperature.
2. Observed to flame continuously for 10 seconds or more inside the furnace. Otherwise the material shall be deemed combustible.

It should be noted that mixtures of organic and inorganic materials have a different behaviour pattern to the individual materials, therefore such combinations must be tested and classified accordingly.

Part 5 Ignitability Test

This test classifies combustible materials as either 'easily ignitable' or 'not easily ignitable'. The test is intended for rigid or semi-rigid building materials but is not suitable for fabrics for which separate tests are available. It identifies easily ignitable materials of low heat contribution of which the performance in the fire propagation test (Part 6) does not necessarily indicate the full hazard. If the specimen to be tested is of laminated construction or has faces of different materials both faces must be tested and the classification is related to the thickness of the specimen and may not be valid for other thicknesses of similar construction.

A lighted gas jet is applied to the centre of the specimen face to be tested for 10 seconds and after removal any subsequent flaming is noted. If the specimen continues to flame for more than 10 seconds after removal of the test flame or if burning extends to the edge of the specimen within 10 seconds it is classed as easily ignitable and this is denoted by the letter 'X'; alternatively it can be classed as not easily ignitable and lettered 'P'.

Part 6 Fire Propagation Test

The scope of this test is to provide a means of comparing the contribution of combustible materials to the growth of fire. This test is incomplete without the addition of a report on classification carried out under Part 5 — Ignitability Test. The specimen is given an index of performance (I) ranging from 0 (non-combustible) to 100 in a descending order relating to a time/temperature curve which in turn is related to the rise in temperature inside the test apparatus over the ambient temperature. The classification 'X' or 'P' from the ignitability test is also given.

Part 7 Surface Spread of Flame Test

Two tests are covered in Part 7, a large-scale test to determine the tendency of materials to support the spread of flame across their surface and to classify this in relation to exposed surfaces of walls and ceilings. The second is a small-scale test intended for preliminary testing, development and quality control purposes. It must be noted that there is no direct correlation between the two tests.

The sample for testing must have any surfacings or coatings applied in the usual manner. The 230 × 900 mm sample is fixed in the holder of the apparatus and subjected to radiant heat from the test furnace. During the first minute of the test a luminous gas flame is applied to the furnace end of the specimen. Throughout the 10 minutes' duration of the test the distance of flame spread over the surface against the time taken is noted and the sample is placed into one of four classes set out in Table 1 of Part 7.

Classification	Flame spread at 1½ minutes		Final flame spread	
	Limit (mm)	Tolerance for one specimen in sample (mm)	Limit (mm)	Tolerance for one specimen in sample (mm)
Class 1	165	25	165	25
Class 2	215	25	455	45
Class 3	265	25	710	75
Class 4		Exceeding Class 3 Limits		

Approved Document B specifies a higher class than class 1 called class 0 which is defined as a non-combustible material throughout or if the surface material is tested in accordance with BS 476 : Part 6 shall have an index (I) not exceeding 12 and a sub-index (i_1) not exceeding 6.

Part 8 Fire Resistance of Elements

The term 'fire resistance' relates to complete elements of construction and not to the individual materials of which elements are composed. The tests enable elements of construction to be assessed according to their ability to retain their stability, to resist the passage of flame and hot gases and to provide the necessary resistance to heat transmission.

The following elements of construction are covered by the tests laid down in Part 8:

1. Load bearing and non-load bearing walls and partitions.
2. Floors.
3. Flat roofs.
4. Columns.
5. Beams.
6. Suspended ceilings protecting steel beams.
7. Door and shutter assemblies.
8. Glazing.

Wherever possible the specimens used in the tests should be full size and be fully representative of the element including at least one of each type of joint.

During the tests the specimen is heated from one side by a furnace which can produce a positive pressure at standard heating conditions until failure occurs or the test is terminated. During the test the following observations are made and noted:

1. *Stability* — deformation of the specimen, occurrence of collapse or any other factor which could affect its stability. Non-load bearing constructions — failure occurs when collapse of the specimen takes place. Load bearing constructions must support their loads during heating period and for 24 hours after heating period; if collapse occurs stability time is 80% of time taken to collapse.
2. *Integrity* — a 100 × 100 mm cotton wool pad is held over the centre of any crack through which flames and gases can pass. The pad is held 30 mm from and parallel to the crack for a period of 10 seconds to determine if hot gases can cause ignition. The observation is repeated at frequent intervals.
3. *Insulation* — unexposed face of elements having a separating function is observed at intervals of not more than 5 minutes. Failure is deemed to occur if (*a*) mean temperature rises more than $140°C$ above initial temperature or (*b*) point temperature rises more than $180°C$ above initial temperature.

The results are given in minutes from the start of the test until failure occurs in one or any of the above observations.

Fire resistance of an element of construction is given in minutes and is the time from the start of the test until failure occurs as given below for each element group.

Walls and Partitions: specimen to be full size or a minimum of 2.500 × 2.500 m and loaded to simulated actual site conditions. Fire resistance — time taken to failure by any one of the three observations.

Floors and flat roofs: specimen to be full size or a minimum of 2.500 m wide × 4.000 m span and loaded to simulate actual site conditions. If ceiling is intended to add to the fire resistance it must be included in the test specimen. Fire resistance — failure time for each observation is noted, specimen is deemed to have failed in stability if deflection exceeds L/30 where L = clear span.

Columns: specimen to be full size or have a minimum length of 3.000 m and loaded to simulate actual site conditions. The specimen is heated on all exposed faces. Fire resistance — time taken for failure in stability only.

Beams: specimen to be full size or have a minimum span of 4.000 m and loaded to simulate actual site conditions. If the beam is exposed to fire on three faces a deck not less than 75 mm thick shall be included in the specimen. Fire resistance — time taken for failure in stability only which is deemed to have occurred if the deflection exceeds L/30 where L = clear span.

Suspended ceilings protecting steel beams: specimen to be full size or a minimum of 2.500 × 4.000 m. The steel beams used to support the specimen have the top flange covered with a lightweight concrete floor at least 130 mm thick. The specimen must be fitted with any light fittings or similar outlets. Fire resistance — limit of effective protection has been deemed to have been reached when:

1. One or more tiles or panels become dislodged or,
2. Loaded beams are unable to support load or mean temperature of beam exceeds $550°C$ or maximum temperature of beam exceeds $650°C$ or deflection exceeds L/30 where L = clear span or,
3. Where an unloaded specimen is used the beam temperature at any point is not less than $400°C$.

Doors: specimen to be full size or a minimum of 2.500 × 2.500 m and complete with any furniture and fittings. The sample is to be fixed in a wall type similar to that expected on site or in a 100 mm thick brick or concrete wall for up to a 2-hour test or a 200 mm thick brick or concrete wall for over a 2-hour test. Fire resistance — specimen to be given a failure time for all three observations. If the cotton wool pad, used in the integrity observation, cannot be used because of radiant heat from the specimen, failure is deemed to have occurred if an unobstructed gap exceeding 6 mm wide × 150 mm long occurs.

Glazing: specimen to be full size or a minimum of 2.500 × 2.500 m and to include fittings and surrounds. If the specimen is not part of a prefabricated system it is to be housed in a brick or concrete wall. Fire resistance — time taken for failure in both stability and integrity are given. Comments made regarding the use of cotton wool pads for doors are also valid for glazing.

The methods and necessity of testing building materials and components under fire conditions are constantly being reviewed and extended. A typical example is the development for measuring the optical density of smoke produced by small specimens of material. The scope of this test is limited to lining materials used in buildings and the measurement of optical density is measured in terms of the transmittance of a parallel beam of light falling on to a photoelectric cell. The results are recorded, computed and the material given a rating, zero being the best performance.

BUILDING REGULATIONS — PART B

One of the major aims of this part of the Building Regulations is to limit the spread of fire and this is achieved by considering the use of a building, fire resistance of structural elements and surface finishes, size of the building or parts of a building and the degree of isolation between buildings or parts of buildings.

Part B of Schedule 1 to the Building Regulations 1985 contains three regulations which are concerned with fire spread under the headings of:

Internal fire spread (surfaces) — Regulation B2
Internal fire spread (structure) — Regulation B3
External fire spread — Regulation B4

Approved Document B which supports these regulations is comprehensive and gives recommendations and guidance as to meeting the performance requirements set out in the actual regulations. It is not proposed to fully analyse each recommendation but to use the recommendations as illustrations as to how the objectives can be achieved.

A full understanding of the terminology used is important to comprehend the recommendations being made in the Approved Document and these are given in Appendix L under the heading of definitions. Figure IV.1 illustrates some of the general definitions given in this appendix. Other interpretations which must be clearly understood are elements of structure, fire stops, relevant and notional boundaries and unprotected areas. These are illustrated in Figs. IV.2, IV.3, IV.4, and IV.5 respectively.

The use of a building enables it to be classified into one of the nine purpose groups given in Approved Document B. The purpose groups can apply to a whole building, a separated part or a compartment of a building. All buildings covered by Part B of the Building Regulations should be included in one of these purpose groups. The nine purpose groups are set out in two main divisions, namely: residential (those with sleeping accommodation) and; non-residential (those without sleeping accommodation). Each purpose group has a descriptive title thus:

Dwellinghouse — does not include a flat or a building containing flats.

Flat — self-contained and includes a maisonette.

Institutional — hospital, home, school or other similar establishment where persons sleep on the premises.

Other Residential — hotel, boarding house, hostel and any other residential purpose not described above.

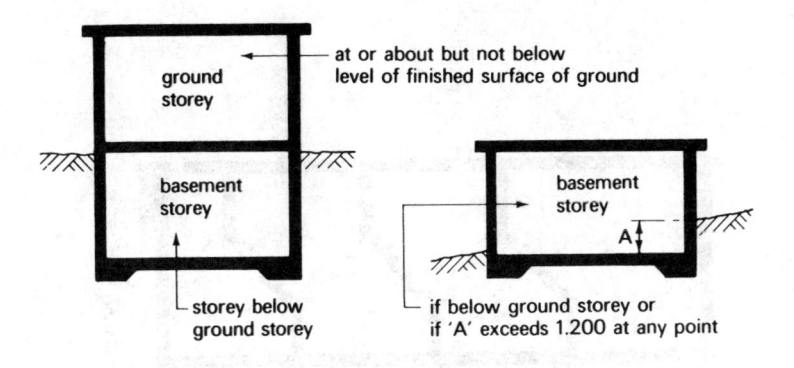

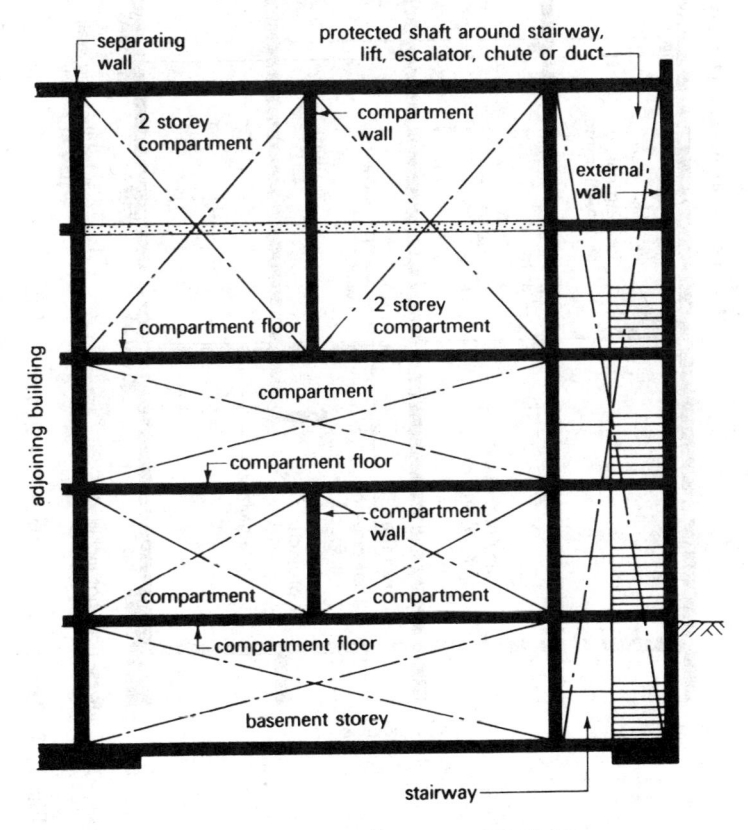

Fig. IV.1 Approved Document B – general definitions

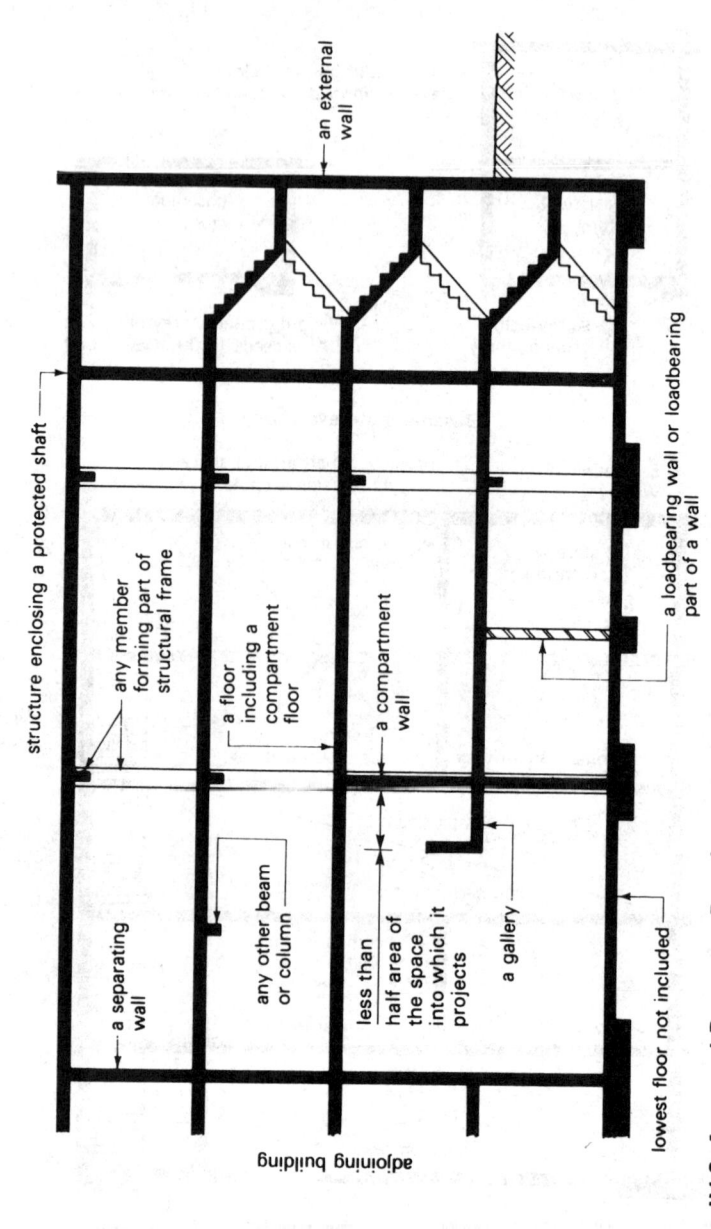

Fig. IV.2 Approved Document B – elements of structure

Appendix L — fire stop means a barrier or seal which would prevent or retard the passage of smoke or flame within a cavity, around a pipe where it passes through a wall or floor or at junctions between elements of structure

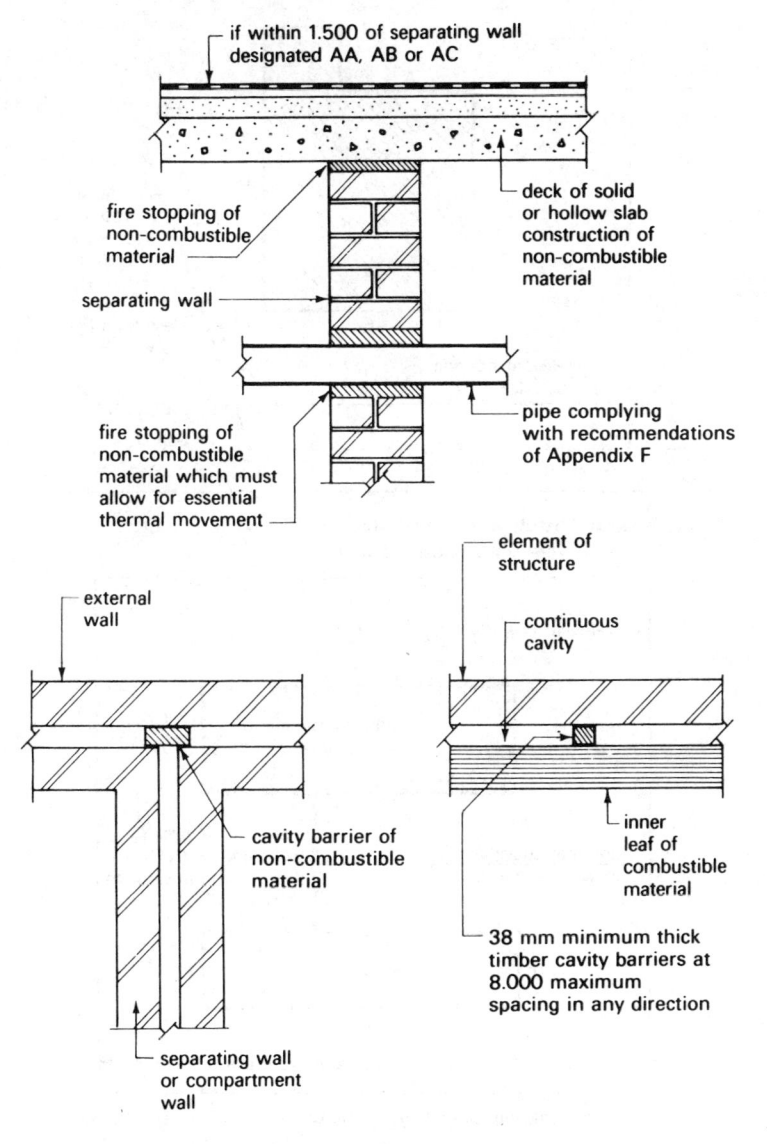

- if within 1.500 of separating wall designated AA, AB or AC
- deck of solid or hollow slab construction of non-combustible material
- fire stopping of non-combustible material
- separating wall
- pipe complying with recommendations of Appendix F
- fire stopping of non-combustible material which must allow for essential thermal movement
- external wall
- cavity barrier of non-combustible material
- separating wall or compartment wall
- element of structure
- continuous cavity
- inner leaf of combustible material
- 38 mm minimum thick timber cavity barriers at 8.000 maximum spacing in any direction

Fig. IV.3 Approved Document B — firestopping

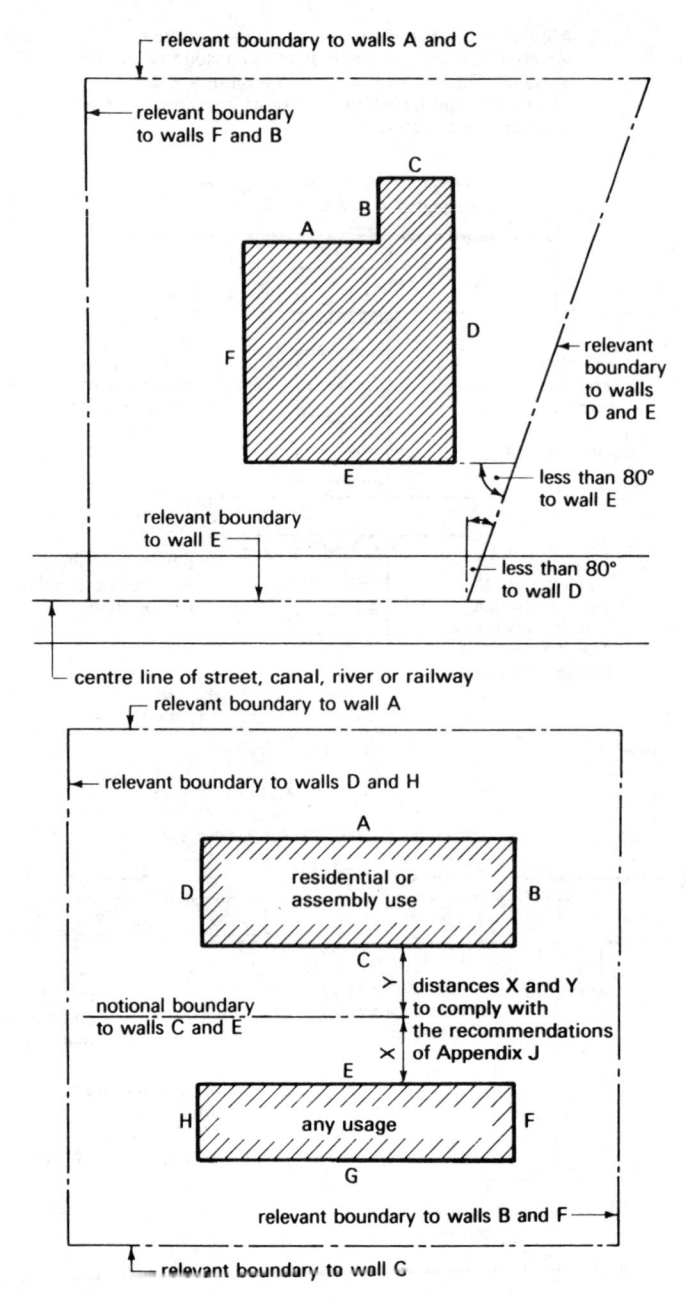

Fig. IV.4 Approved Document B — boundaries

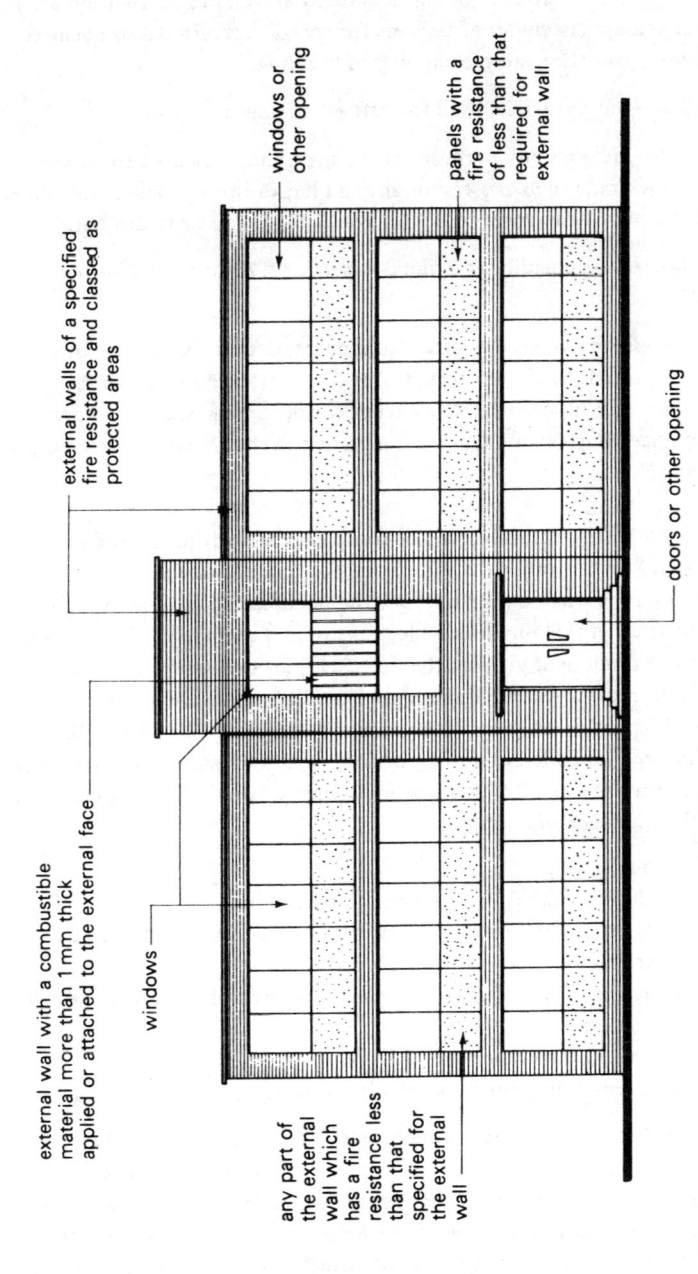

Fig. IV.5 Approved Document B – unprotected areas

Assembly — a public building as defined in Building Regulation 2(2) or a place of assembly of persons for social, recreational or business but not an office, shop or industrial building.

Office — any premises used for office and administration work.

Shop — includes any premises not being a shop but used for any form of retail trade such as a restaurant and hairdressers or where members of the public can enter to deliver goods for repair or treatment.

Industrial — generally as defined in section 175 of the Factories Act 1961.

Other Non-residential — place for storage, deposit or parking of goods and materials (including vehicles) and any other non-residential purpose not described above except for detached garages and carports not exceeding 40 m^2 which are included in the Dwellinghouse purpose group.

The use of fire-resistant cells or compartments within a building (Fig. IV.1) are a means of confining an outbreak of fire to the site of origin for a reasonable time to allow the occupants a chance to escape and the fire-fighters time to tackle, control and extinguish the fire. The Approved Document gives a table for each purpose group setting out the maximum recommended dimensions for buildings or compartments in terms of height of building, floor area and cubic capacity. The tables also give the recommended minimum periods of fire resistance for all elements of structure (see Fig. IV.2). These minimum periods of fire resistance are given in hours according to:

1. Purpose group.
2. Height of building or of separated part in metres.
3. Floor area of each storey or each storey within a compartment in square metres.
4. Cubic capacity of building or cubic capacity of a compartment in cubic metres.
5. Ground or upper storey.
6. Basement storey including the floor over.

The recommended minimum periods of fire resistance given in the tables are of prime concern to the designer and contractor. These tables do not state how these minimum periods are to be achieved but reference is made to the current edition of the BRE report *Guidelines for the construction of fire resisting elements* which gives appropriate and common methods of construction for the various notional periods of fire

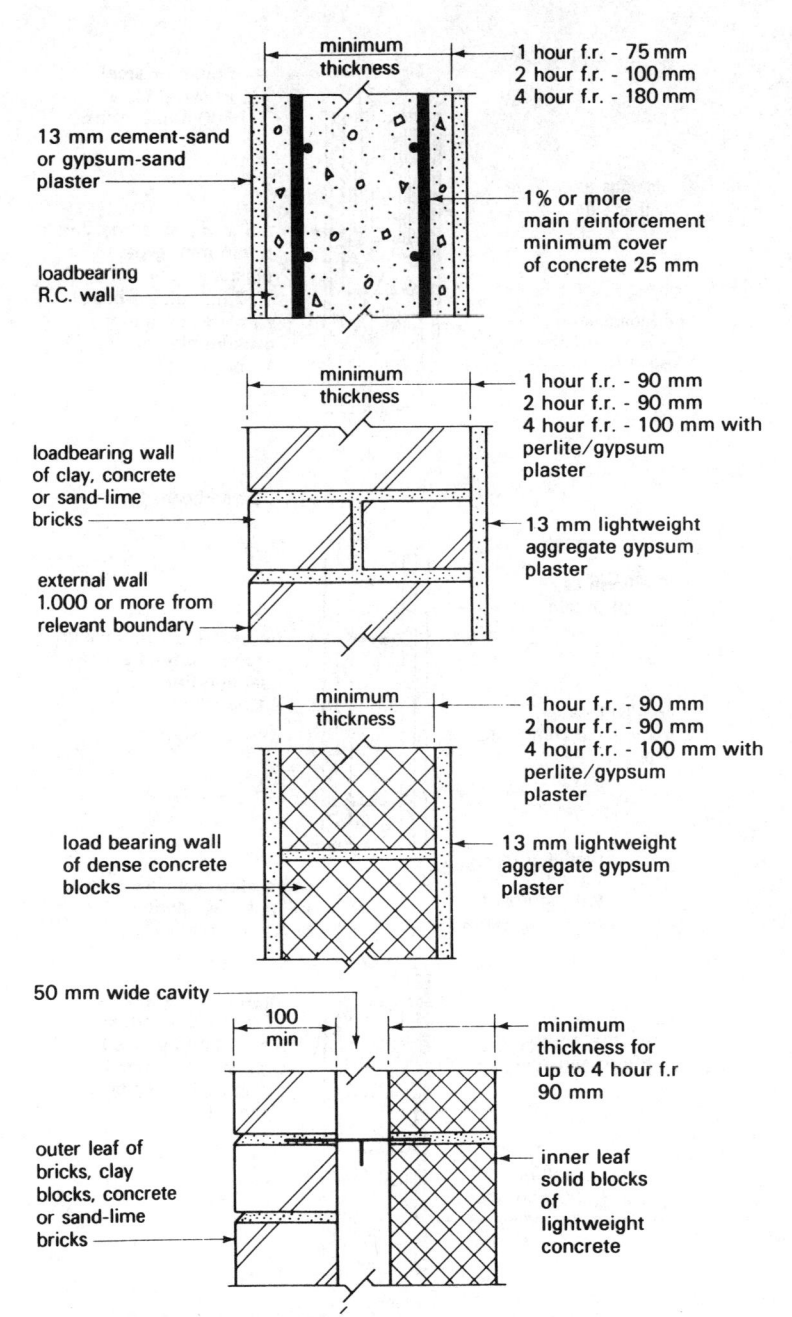

Fig IV.6 Fire Resistance—walls of masonry construction

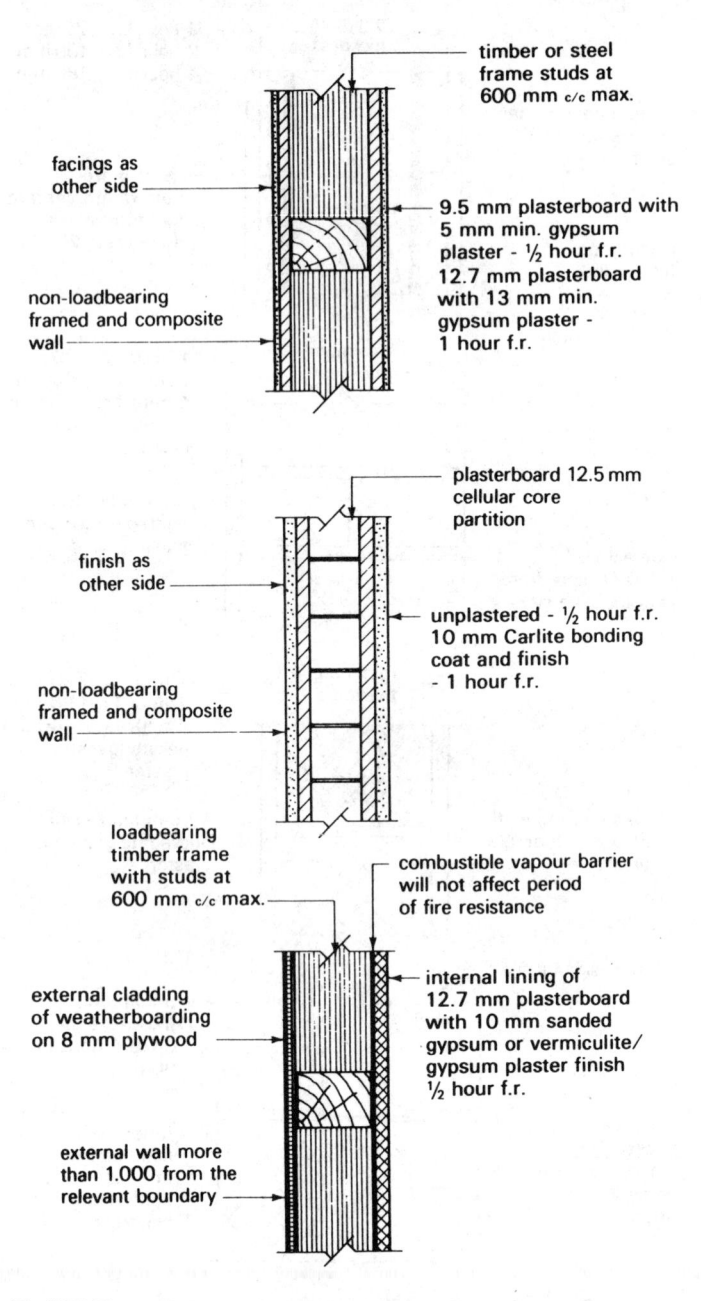

Fig IV.7 Fire resistance—framed and composite walls

free standing R.C. column

minimum dimension

unplastered:-
1 hour f.r. - 200 mm
2 hour f.r. - 300 mm
4 hour f.r. - 450 mm
actual cover over reinforcement:
1 hour f.r. - 25 mm
2 hour f.r. - 35 mm
4 hour f.r. - 35 mm

binders

main reinforcement

R.C. column built in wall

separating or compartment wall

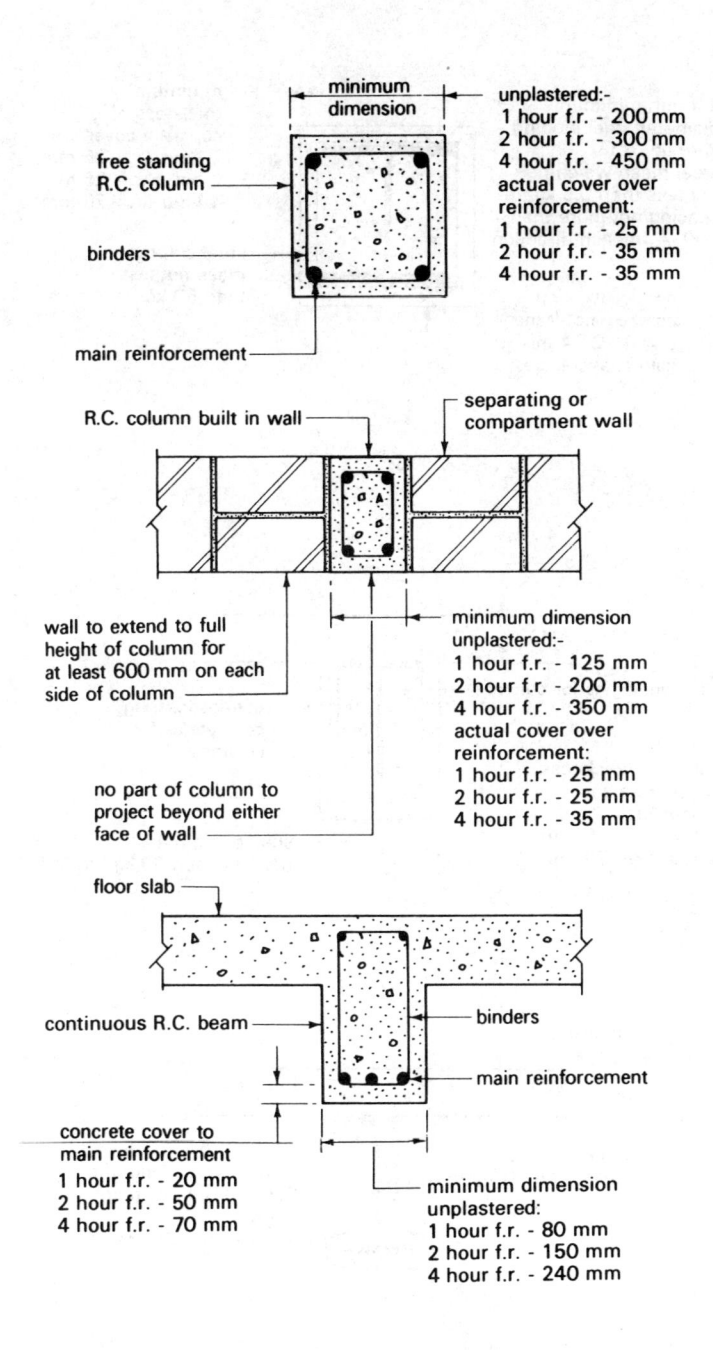

wall to extend to full height of column for at least 600 mm on each side of column

no part of column to project beyond either face of wall

minimum dimension unplastered:-
1 hour f.r. - 125 mm
2 hour f.r. - 200 mm
4 hour f.r. - 350 mm
actual cover over reinforcement:
1 hour f.r. - 25 mm
2 hour f.r. - 25 mm
4 hour f.r. - 35 mm

floor slab

continuous R.C. beam

binders

main reinforcement

concrete cover to main reinforcement
1 hour f.r. - 20 mm
2 hour f.r. - 50 mm
4 hour f.r. - 70 mm

minimum dimension unplastered:
1 hour f.r. - 80 mm
2 hour f.r. - 150 mm
4 hour f.r. - 240 mm

Fig IV.8 Fire resistance—R.C. columns and beams

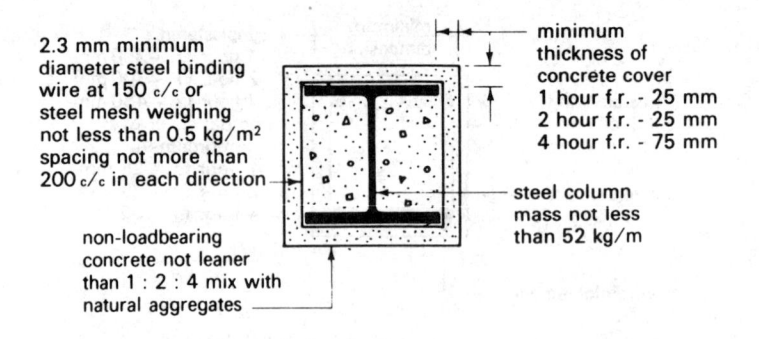

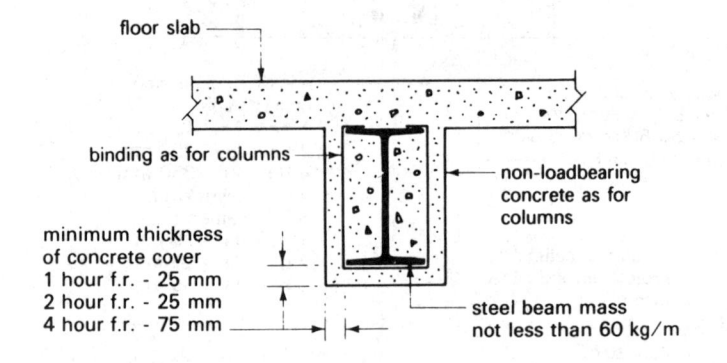

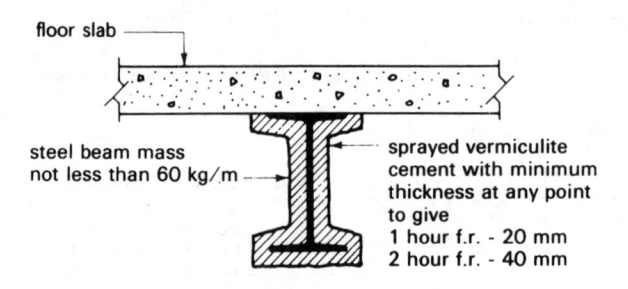

Fig IV.9 Fire resistance—steel columns and beams

all hollow protection to be effectively
sealed at each floor level

unplastered solid bricks of clay, composition or sand-lime reinforced in every horizontal joint

minimum thickness of brickwork
1 hour f.r. - 50 mm
2 hour f.r. - 50 mm
4 hour f.r. - 100 mm

steel column mass not less than 52 kg/m

metal lathing with angle beads to corners for 2 hour f.r.

metal lathing with trowelled lightweight aggregate gypsum plaster

minimum thickness of plaster cover
1 hour f.r. - 13 mm
2 hour f.r. - 20 mm

steel column mass not less than 52 kg/m

vermiculite/gypsum plaster

1.6 mm wire binding at 100 mm centres

plasterboard of required thickness

steel column mass not less than 52 kg/m

minimum thickness of plasterboard and plaster
1 hour f.r. - 9.5 mm plasterboard and 10 mm plaster or 19 mm plasterboard and 10 mm plaster
2 hour f.r. - 19 mm plasterboard and 20 mm plaster

insulating boards screwed to 50 × 25 asbestos battens

asbestos insulating boards of density 500-900 kg/m^3

minimum thickness of insulating board
1 hour f.r. - 12 mm
1½ hour f.r. - 19 mm

steel column mass not less than 52 kg/m

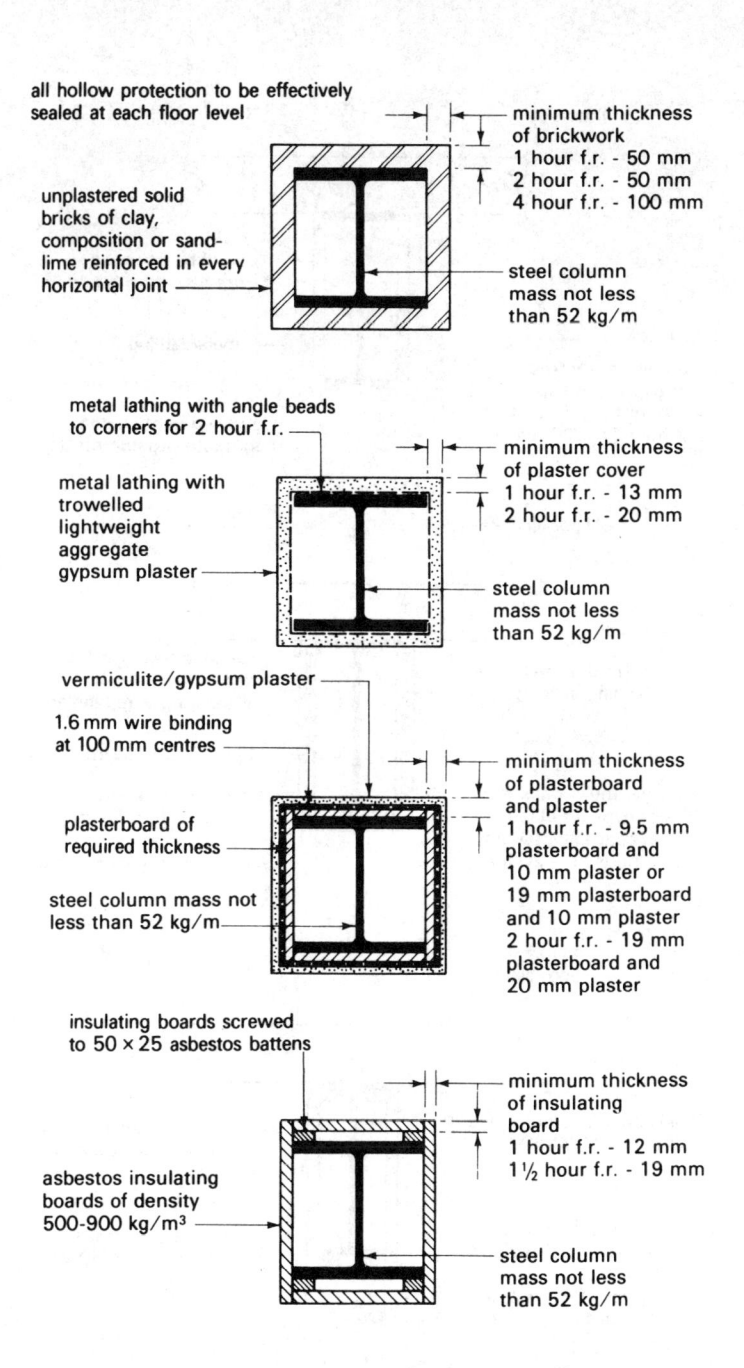

Fig IV.10 Fire resistance - hollow protection to steel columns

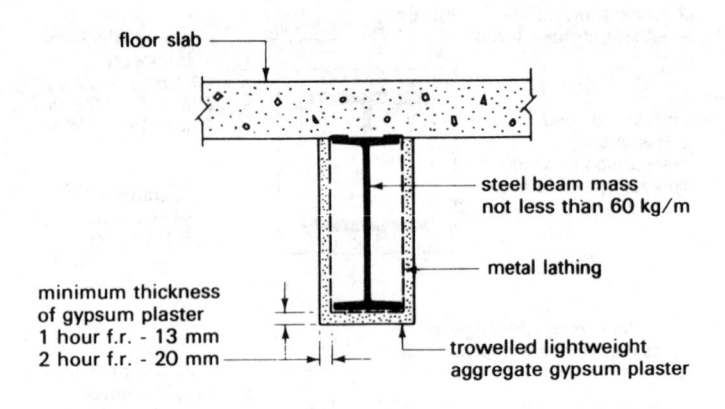

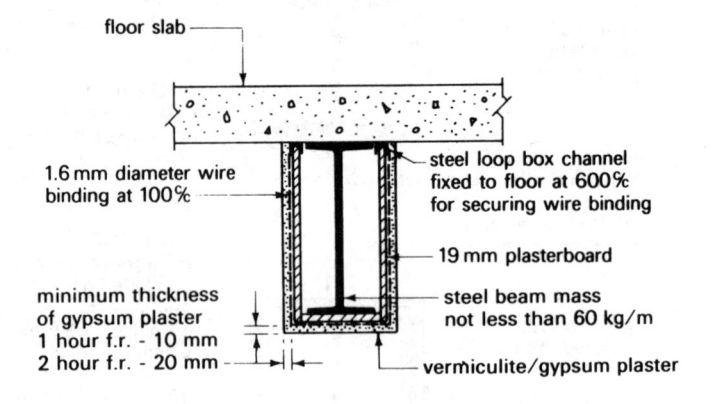

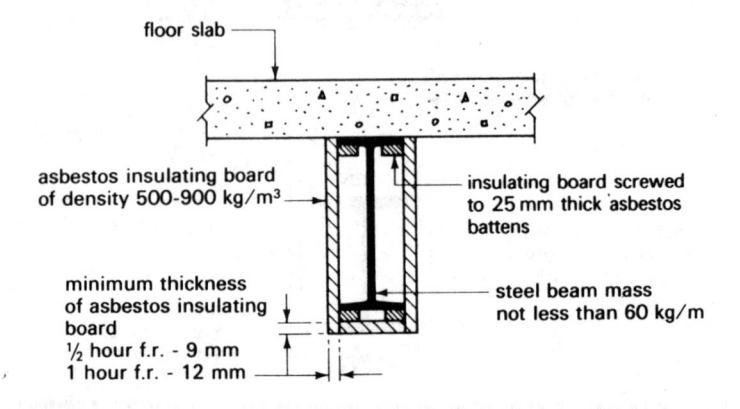

Fig IV.11 Fire resistance - hollow protection to steel beams

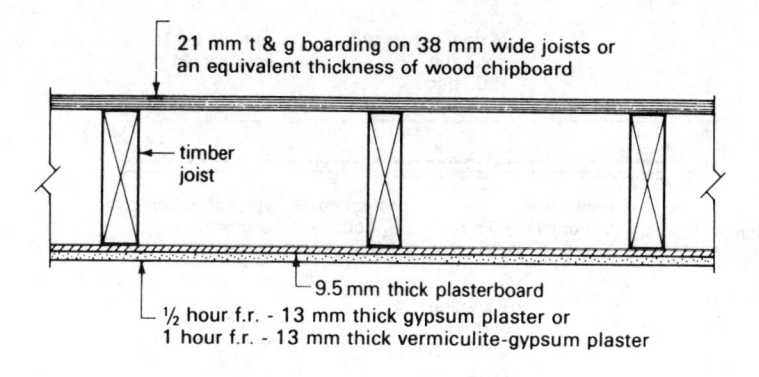

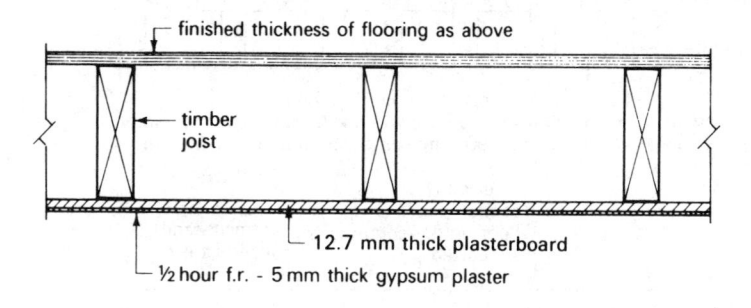

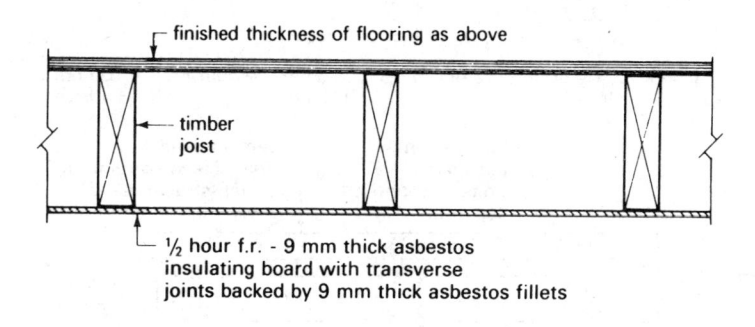

Fig IV.12 Fire resistance - timber floors

111

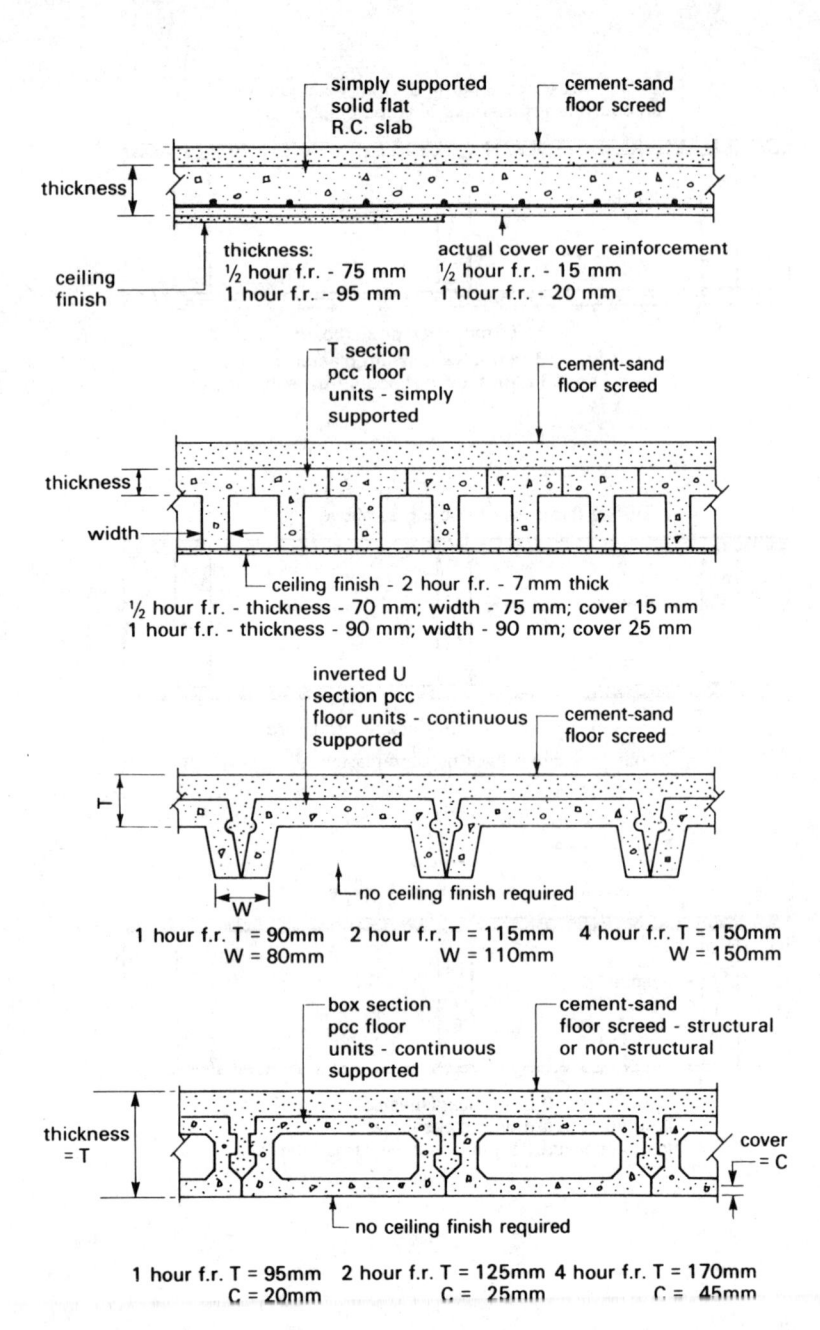

Fig IV.13 Fire resistance - concrete floors

Simply supported solid flat R.C. slab with cement-sand floor screed, ceiling finish:
- thickness: ½ hour f.r. - 75 mm; 1 hour f.r. - 95 mm
- actual cover over reinforcement: ½ hour f.r. - 15 mm; 1 hour f.r. - 20 mm

T section pcc floor units - simply supported with cement-sand floor screed:
- ceiling finish - 2 hour f.r. - 7 mm thick
- ½ hour f.r. - thickness - 70 mm; width - 75 mm; cover 15 mm
- 1 hour f.r. - thickness - 90 mm; width - 90 mm; cover 25 mm

Inverted U section pcc floor units - continuous supported with cement-sand floor screed; no ceiling finish required:
- 1 hour f.r. T = 90mm, W = 80mm
- 2 hour f.r. T = 115mm, W = 110mm
- 4 hour f.r. T = 150mm, W = 150mm

Box section pcc floor units - continuous supported with cement-sand floor screed - structural or non-structural; thickness = T, cover = C; no ceiling finish required:
- 1 hour f.r. T = 95mm, C = 20mm
- 2 hour f.r. T = 125mm, C = 25mm
- 4 hour f.r. T = 170mm, C = 45mm

resistance for walls, beams, columns and floors. The guidelines are written and presented in a tabulated format which needs to be translated into working details. Figs. IV.6 to IV.13 show typical examples taken from this document and other sources of reference such as manufacturer's data.

Students are encouraged to study manufacturers' literature on the many patent ready-cut, easy-to-fix fire protection systems for standard structural members and other methods such as the use of intumescent paints and materials which expand to form a thick insulating coating or strip on being heated by fire.

The degree of isolation needed between buildings to limit the spread of fire by radiation is covered in Appendix J of Approved Document B. This deals with external walls and, in particular, unprotected areas permitted in relationship to the distance of the building or compartment within the building from the relevant boundary and is applicable to buildings which are not less than 1 metre at any point from the relevant boundary. Appendix J gives three methods for satisfying the requirements of Building Regulation B4 (External Fire Spread). The first is concerned with small residential buildings not included in the Institutional group. A simple table gives the relationship between boundary distance, maximum length of building side and the maximum total of unprotected area permitted.

The second method, which can be used for all buildings, is known as the 'enclosing rectangle method' which can be used to ascertain the maximum unprotected area for a given boundary position or to find the nearest position of the boundary for a given building design. The method consists of placing an enclosing rectangle around the unprotected areas, noting the width and height of the enclosing rectangle and calculating the unprotected percentage in terms of the enclosing rectangle. This information will enable the distance from the boundary to be read direct from the tables given in Appendix J of Approved Document B. It should be noted that in compartmentated buildings the enclosing rectangle is taken for each compartment and not for the whole facade. Typical examples are shown in Fig. IV.14.

Method three is an alternative to the second method described above. The principle of isolation is still followed but by a more precise method which will involve more investigation than reference to an enclosing rectangle. Reference is made to an aggregate notional area which is calculated by taking the sum of each relevant unprotected area and multiplying by a factor given in Table J3 of Appendix J. Which unprotected areas are relevant is given in the appendix. The method entails dividing the relevant boundary into a series of 3.000 spacings called vertical datum and projecting from each point a datum line to

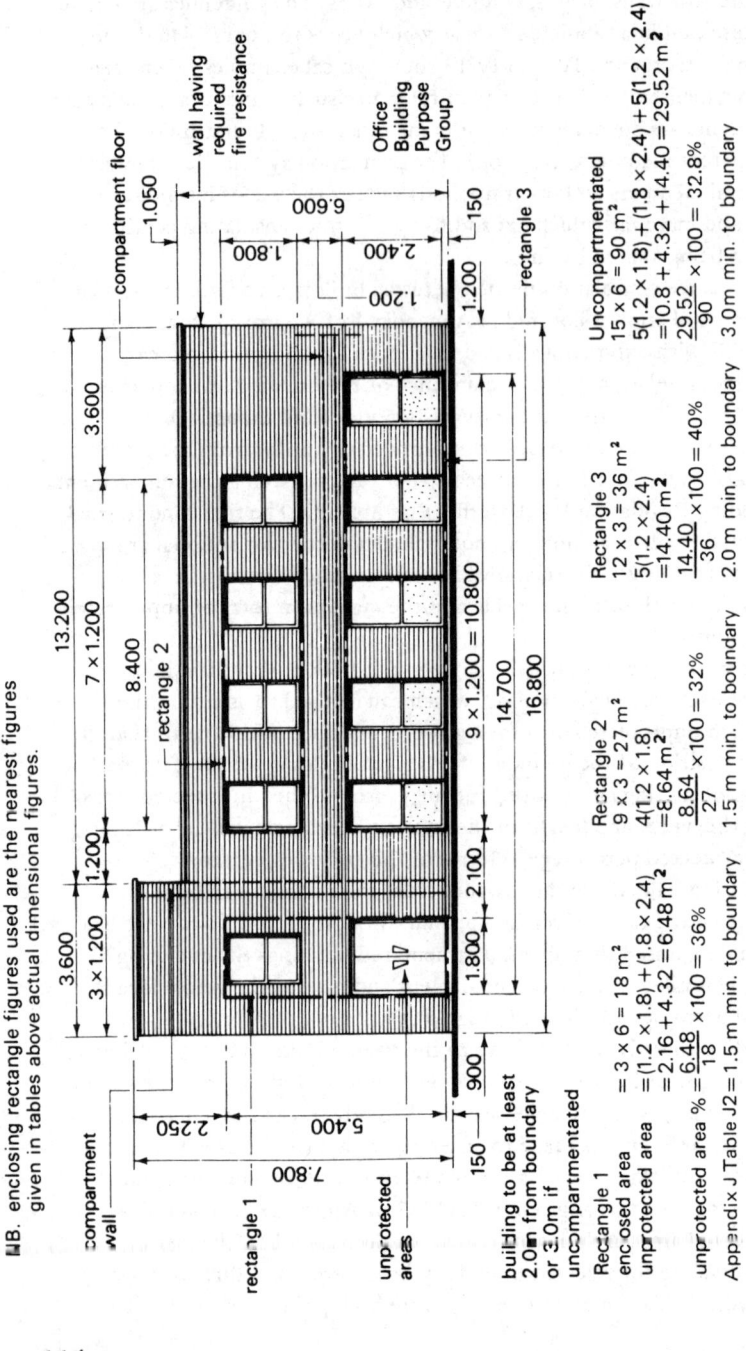

NB. enclosing rectangle figures used are the nearest figures given in tables above actual dimensional figures.

Rectangle 1
enclosed area $= 3 \times 6 = 18 \text{ m}^2$
unprotected area $= (1.2 \times 1.8) + (1.8 \times 2.4)$
$= 2.16 + 4.32 = 6.48 \text{ m}^2$
unprotected area % $= \dfrac{6.48}{18} \times 100 = 36\%$
Appendix J Table J2 = 1.5 m min. to boundary

Rectangle 2
$9 \times 3 = 27 \text{ m}^2$
$4(1.2 \times 1.8)$
$= 8.64 \text{ m}^2$
$\dfrac{8.64}{27} \times 100 = 32\%$
1.5 m min. to boundary

Rectangle 3
$12 \times 3 = 36 \text{ m}^2$
$5(1.2 \times 2.4)$
$= 14.40 \text{ m}^2$
$\dfrac{14.40}{36} \times 100 = 40\%$
2.0 m min. to boundary

Uncompartmented
$15 \times 6 = 90 \text{ m}^2$
$5(1.2 \times 1.8) + (1.8 \times 2.4) + 5(1.2 \times 2.4)$
$= 10.8 + 4.32 + 14.40 = 29.52 \text{ m}^2$
$\dfrac{29.52}{90} \times 100 = 32.8\%$
3.0 m min. to boundary

building to be at least 2.0m from boundary or 3.0m if uncompartmentated

Office Building Purpose Group

Fig V.14 Unprotected areas - enclosing rectangle method

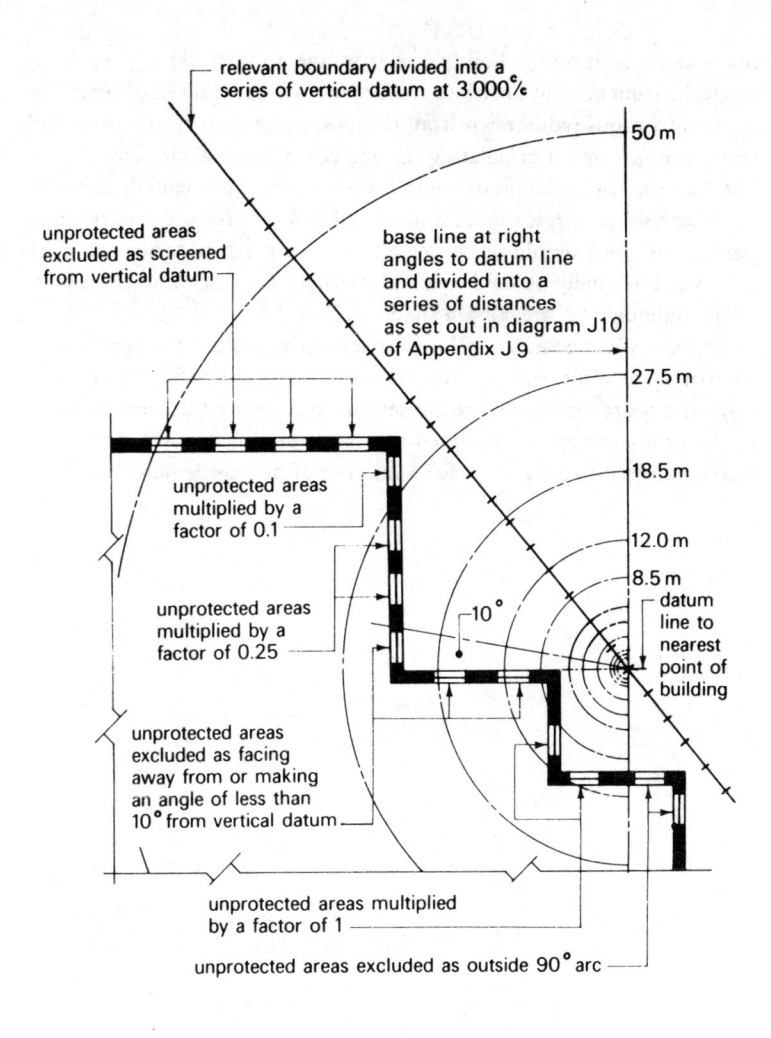

relevant boundary divided into a series of vertical datum at 3.000%

50 m

unprotected areas excluded as screened from vertical datum

base line at right angles to datum line and divided into a series of distances as set out in diagram J10 of Appendix J 9

27.5 m

18.5 m

unprotected areas multiplied by a factor of 0.1

12.0 m

8.5 m

unprotected areas multiplied by a factor of 0.25

10°

datum line to nearest point of building

unprotected areas excluded as facing away from or making an angle of less than 10° from vertical datum

unprotected areas multiplied by a factor of 1

unprotected areas excluded as outside 90° arc

Notes:- 1. Procedure repeated for each vertical datum position.

2. Calculations carried out for any side of a building or compartment.

3. For each position aggregate notional area of the unprotected areas must not exceed:
210 m for residential, assembly or office use.
90 mm for all other purpose groups.

Fig IV.15 Unprotected areas - aggregate notional area method

115

the nearest point on the building. A base line is drawn through each vertical datum at right angles to the datum line and a series of semi-circles of various radii drawn from this point represent the distance and hence the factors for calculating the aggregate notional area (see Fig. IV.15). For buildings or compartments with a residential, assembly or office use the result should not exceed 210 m^2; for other purpose groups the result should not exceed 90 m^2. In practical terms this method would not normally be used unless the situation is critical or the outline of the building was irregular in shape.

The specific recommendations for separating walls, openings in compartment walls, compartment floors, protective shafts, doors, stairways and protecting structure are set out in Approved Document B. Each purpose group is considered separately although many of the recommendations are similar for a number of purpose groups.

12
Means of escape in case of fire

Means of escape in case of fire is concerned with the personal hazard by giving people within a building where an outbreak of fire occurs a reasonable chance to reach an area of safety, taking into account factors such as the risks to human life, unfamiliarity with building layout, problems of smoke and the short space of time available to evacuate the premises before the problems become almost insurmountable.

Fear is a natural response of humans when confronted with uncontrolled fire and in particular fear of smoke, which is justified by the fact that more deaths are caused by smoke and heated gases than by burns. Statistics show that on average approximately 54% of deaths in fires are caused by smoke, 40% by burns and scalds and 6% by other causes. BS 4422 : Part I (Glossary of Terms Associated with Fire) defines smoke as a visible airborne cloud of fine particles, the products of incomplete combustion. Whilst the above statement is indeed true smoke can also be caused by the release into the air of a variety of chemical compounds. The main dangers, contained in smoke, to human life are the carbon dioxide and carbon monoxide gases which are normal products of combustion.

The presence of these gases does not always cause the greatest hazard in human terms since the density of smoke is more likely to create fear than the undetectable gases. Buoyant and mobile dense smoke will spread rapidly within a building or compartment during a fire, masking or even obliterating exit signs and directions. Gases, other than those mentioned previously, are generally irritants which can affect the eyes, causing watering which further impairs the vision, and can also affect the respiratory organs, causing reactions to slow and a loss of directional sense. It is worth remembering that smoke, being less dense than air, rises and that to take up a position as near to the floor as possible will increase the chances of escape.

Carbon dioxide has no smell and is always present in the atmosphere but since it is a product of combustion its volume increases at the expense of oxygen during a fire. The gas is not poisonous but can cause death by asphyxia. The normal amount of oxygen present in the air is approximately 21%; if this is reduced to 12% abnormal fatigue can be experienced; down to about 6% it can cause nausea, vomiting and loss of consciousness; below 6% respiration is difficult, which can result in death. Carbon dioxide will not support combustion and can cause a fire to be extinguished if the content by volume exceeds 14%, a fact used by fire fighters in their efforts to deal with an outbreak of fire.

Carbon monoxide, like carbon dioxide, is odourless but it is very poisonous and, having approximately the same density as air, will spread rapidly. A very small concentration (0.2% by volume) of this colourless gas can cause death in about 40 minutes. The first effects are dizziness and headaches followed in 5 to 10 minutes by loss of consciousness leading to death. As the concentration increases so the time lapse from the initial dizziness to death decreases so that by the time the concentration has reached about 1.3% by volume death can take place within a minute or two.

The heat which is associated with fire and smoke can also be injurious and even fatal. Temperatures in excess of $100°C$ can cause damage to the windpipe and lungs resulting in death within 30 minutes or sooner as the temperature rises. An interesting fact which emerges from statistics is that females generally have longer survival periods than males, and as would be expected the survival time decreases with age. Injuries caused by heat are generally in the form of burns followed very often by shock which can be fatal in many cases.

The above has been written not to frighten, but to emphasise the necessity for an adequate means of escape to give occupants and visitors a reasonable chance to reach an area of safety should a fire occur. To this end a maze of legislation exists to guide the designer in his task of planning escape routes without being too restrictive on the overall design concept. Scotland has its own regulations and byelaws whilst the rest of the country is covered by Part B of the Building Regulations and various Acts of Parliament.

BUILDING REGULATIONS 1985

Building Regulation B1 requires that in case of fire a means of escape leading from the building to a place of safety outside the building must be capable of being safely and effectively used at all times. The regulation applies only to certain types of buildings, namely: dwellinghouses of three or more storeys; flats of three or more storeys; offices and shops. The regulation is supported by a document entitled *The Building Regulations 1985 — Mandatory rules for means of escape in case of fire* and is the only method of complying with Building Regulation B1 unless the local authority agree upon a relaxation. Most other types of building are covered by designations made under the Fire Precautions Act 1971.

Planning Escape Routes

When escape routes are being planned the type of person likely to be involved must be considered. Occupants of flats will be familiar with the layout of the premises whereas customers in a shop may be completely unfamiliar with their surroundings. In schools the fundamental principle is the provision of an alternative means of escape and in hospitals the main concern is with the adequacy of the means of escape from all parts of the building.

In the context of means of escape in case of fire the building and its contents are of secondary importance. The provision of a safe escape route should, however, allow at the same time an easy access for the fire brigade using the same routes, and since these routes are protected the risk of fire spread is minimised. In practice the provision of an adequate means of escape and structural fire protection of the building and its contents are virtually inseparable. Each building has to be considered as an individual exercise but certain common factors prevail in all cases:

1. An outbreak of fire does not necessarily imply the evacuation of the entire building.
2. Rescue facilities of the local fire brigade should not be considered as part of the planning of means of escape.
3. Persons should be able to reach safety without assistance when using the protected escape routes.
4. All possible sources of an outbreak and the course the fire is likely to take should be examined and the escape routes planned accordingly.

FIRE PRECAUTIONS ACT 1971

This Act received the Royal Assent on 27 May 1971 and is designed to make further provisions for the protection of persons from fire risks. The provisions of this Act do not apply to Northern Ireland and section 11, which deals with means of escape, does not apply to the old Greater London Council area, which has its own byelaws, or to Scotland, which has its own regulations.

A fire certificate is required on any premises designated by the Minister of State at the Home Office within the following classes of use:

1. Sleeping accommodation.
2. Institutions providing treatment or care.
3. Entertainment, recreation or instruction.
4. Teaching, training or research.
5. Any purpose involving access to the premises by members of the public by payment or otherwise.

Certain types of buildings and premises are exempted from the provisions of this Act and these can be listed as follows:

1. Premises covered by the Offices, Shops and Railway Premises Act 1963.
2. Premises covered by the Factories Act.
3. A quarry or mine.
4. Churches, chapels and places of worship.
5. Houses occupied as a single dwelling.

In certain circumstances the fire authority may make it compulsory to have a fire certificate in particular where the premises have a room used as living accommodation which is:

1. Below the ground floor of a building.
2. Two or more floors above the ground.
3. A room of which the floor is 6.000 m or more above the surface of the ground on any side of the building.

The Secretary of State for the Environment has power, by virtue of sections 4 and 6 of the Public Health Act 1961, section 11 of the Fire Precautions Act 1971 and the Building Act 1984 to make building regulations with regard to means of escape in case of fire.

Applications for a fire certificate are made to the Fire Authority on form FP1. The Fire Authority is as defined in the Fire Services Act 1947 and usually the name and address of the relevant Fire Authority can be obtained from the local authority or council offices. It is generally the occupiers' responsibility to make the application and submit any plans, specifications and details required. Penalties for offences under this Act range from £50 to £400 fines plus two years' imprisonment.

The first Designating Order was made on 21 February 1972 and applies to all existing hotels and boarding houses with accommodation for over six people whether guests or staff unless where accommodation is for less than six people and such accommodation is above first floor or below ground floor.

The basic principles embodied in the Act with regards to means of escape in case of fire are:

1. Limitation of travel distances.
2. Escape route considered in 3 stages
   (a) travel distance within rooms;
   (b) travel distance from rooms to a stairway or final exit;
   (c) travel within stairways and to final exit.
3. Provision of a protected route which is defined as a route for persons escaping from fire which is separated from the remainder of the building by fire-resisting doors (except doors to lavatories), fire resistant walls, partitions and floors.

The basic design principles relating to the 3 stages of escape are shown in Figs. IV.16, IV.17 and IV.18.

Flats and Maisonettes

At one time only accommodation over 24.000 m above ground level would have been considered, this being the height over which external rescue by the fire brigade was impracticable. It has become apparent that even with dwellings within reach of fire fighters' ladders external rescue is not always possible. This is because present-day traffic conditions and congestion may prevent the appliance from approaching close to the building or may delay the arrival of the fire brigade.

As with other forms of buildings the only sound basis for planning means of escape from flats and maisonettes is to identify the positions of all possible sources of any outbreak of fire and to predict the likely course the fire, smoke and gases would follow. The planning should be considered in 3 stages:

1. Risk to occupants of the dwelling in which the fire originates.
2. Risk to occupants of adjoining dwellings should the fire or smoke penetrate the horizontal escape route or common corridor.
3. Risk to occupants above the level of the outbreak particularly on the floor immediately above the source of the fire.

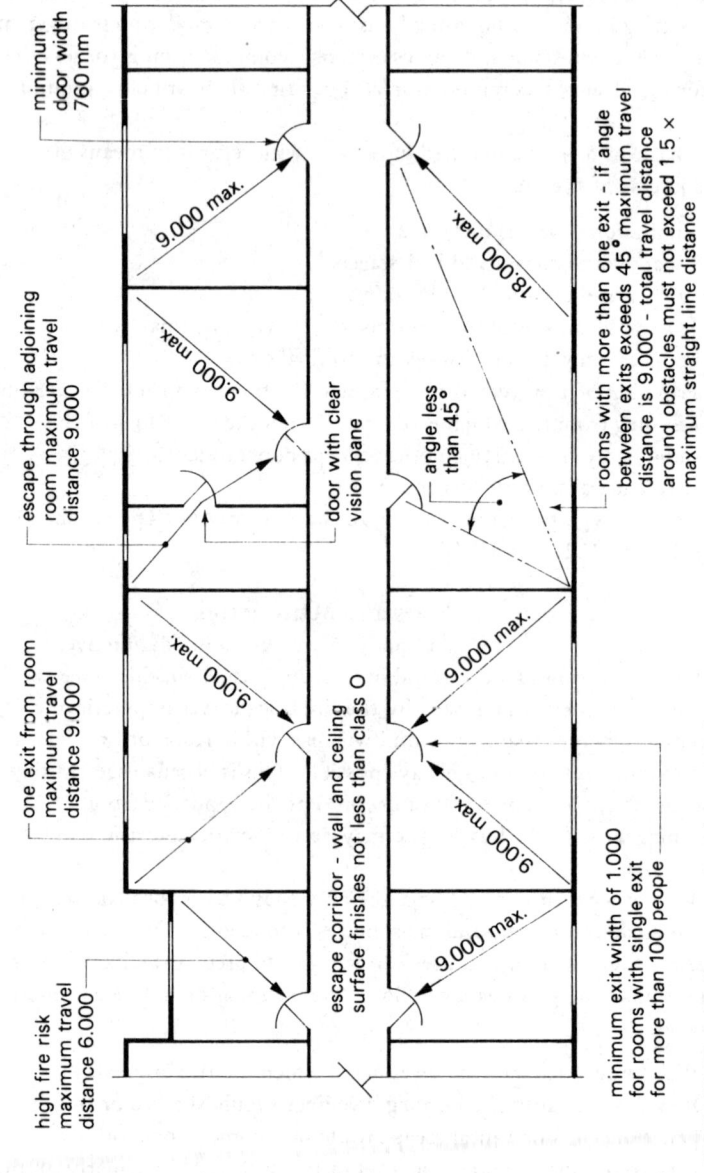

Fig IV.16 Means of escape - hotels - stage 1

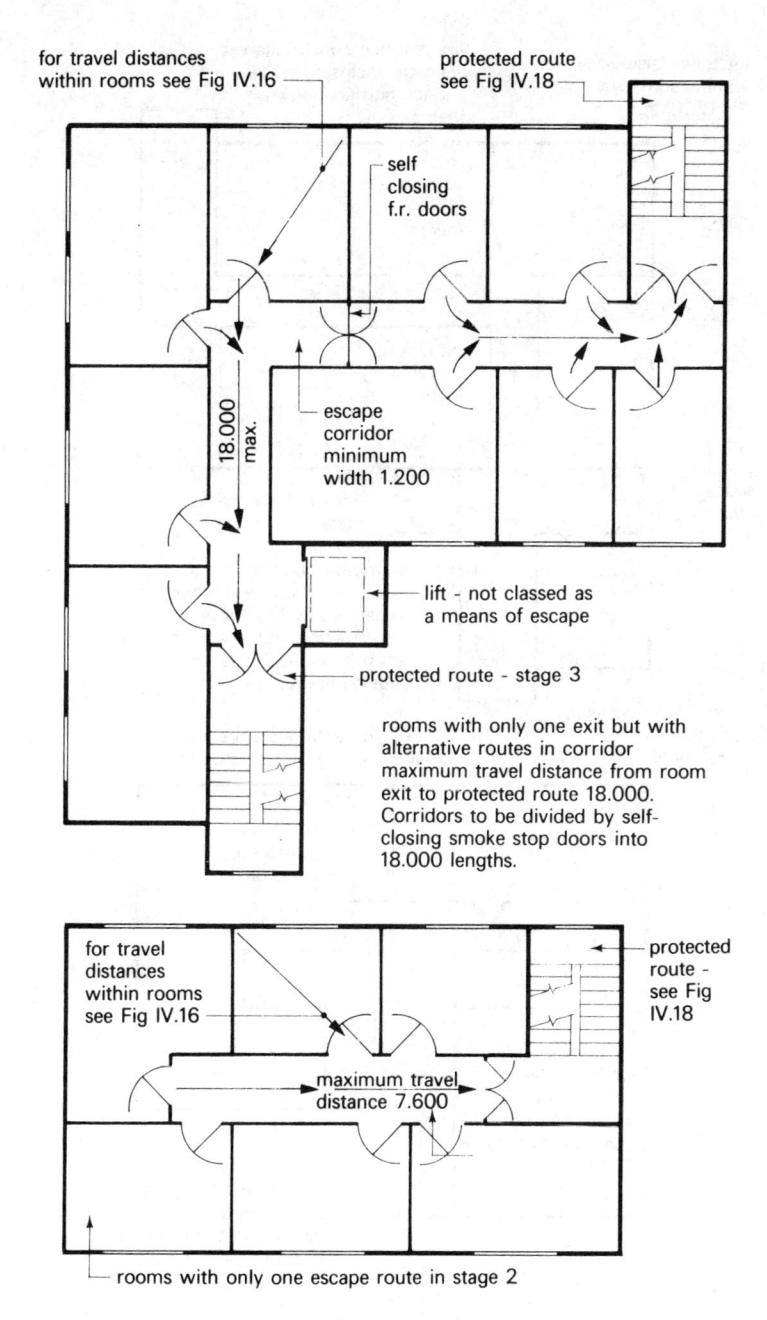

Fig IV.17 Means of escape - hotels - stage 2

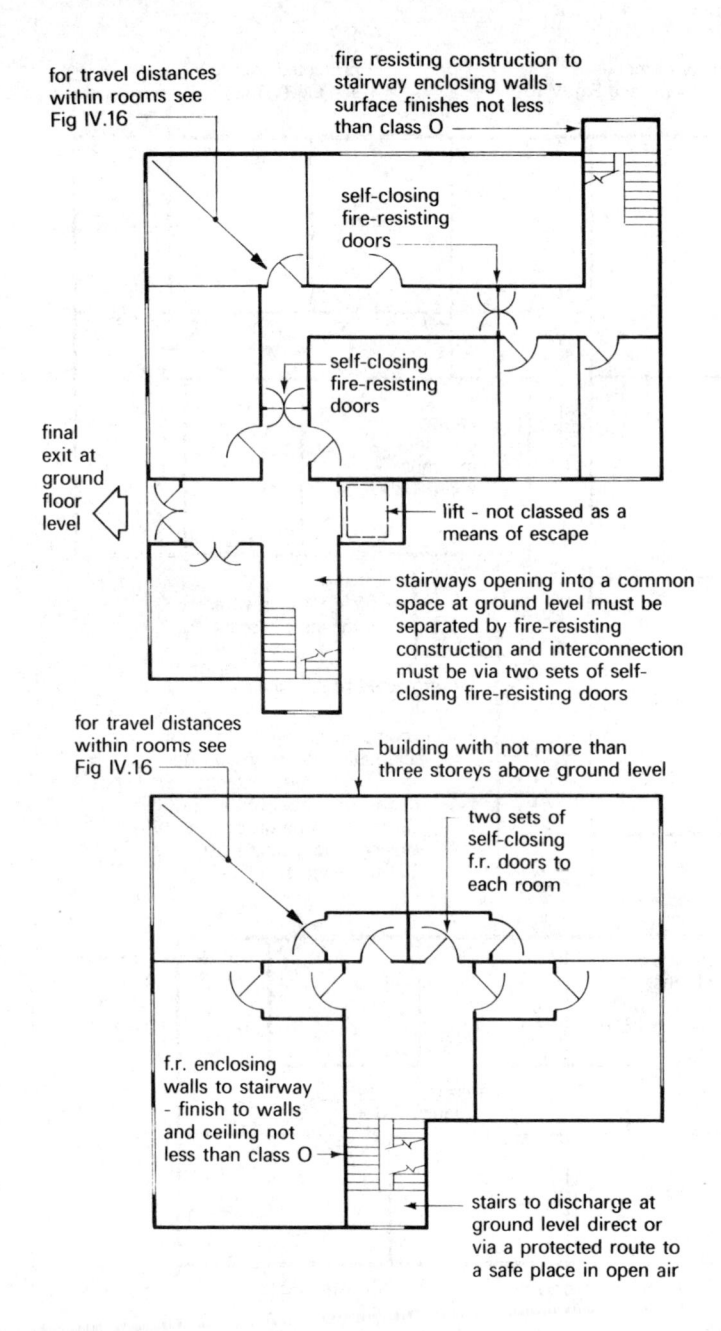

Fig IV.18 Means of escape — hotels — stage 3

Stage 1: most of the serious accidents and deaths occur in the room in which the outbreak originates. Fires in bedrooms have increased in the last decade due mainly to the increased use of electrical applicances such as heated blankets which have been improperly maintained or wired. Social habits, such as watching television, has enabled fires starting in other rooms to develop to a greater extent before detection. Fires occurring in the circulation spaces such as halls and corridors are a serious hazard and the use of paraffin heaters in these areas should be discouraged. Risks to occupants in maisonettes are higher than those incurred in flats since fire and smoke will spread more rapidly in the vertical direction than in the horizontal direction.

Stage 2: concerned mainly with the safety of occupants using the horizontal escape route where the major aim is to ensure that should a fire start in any one dwelling it will not adversely affect or obstruct the escape of occupants of any other dwelling on the same or adjoining floor.

Stage 3: concerned with occupants using a vertical escape route which means in fact a stairway, for in this context lifts are not considered as a means of escape because:

1. Time delay in lift answering call.
2. Limited capacity of lift.
3. Possible failure of the electricity supply in the event of a fire.

The main objective is to remove the risk of fire or smoke entering a stairway and rendering it impassable above that point. This objective can be achieved by taking the following protective measures:

1. Where there is more than one stairway serving a corridor without cross ventilation there should be smoke-stop doors across the corridor between the doors in the enclosing walls of the stairways to ensure that both stairs are not put at risk in the event of a fire — see Fig. IV.19.
2. Where there is only one stairway serving a corridor without cross ventilation there must be a smoke-stop door between the door in the enclosing walls of the stairway and any door covering a potential source of fire, and the lobby so formed should be permanently ventilated through an adjoining external wall — see Fig. IV.19.
3. Where corridors are provided with cross ventilation by means of any mechanical device a smoke-stop door is recommended only for the enclosing wall separating the stairway from the corridor.

The basic means of escape requirements for flats can be enumerated thus:

1. Every flat should have a protected entrance hall of half-hour fire resistance.
2. Every living room should have an exit into the protected entrance hall.
3. Bedrooms should be nearer to the entrance door than the living rooms or kitchen.
4. Doors opening on to a protected entrance hall to be type 3 (see section on doors in this chapter).
5. Maximum travel distance from any bedroom exit to entrance door to be 7.500 m; if exceeded an alternative route is to be provided.

A typical flat example is shown in Fig. VI.20.

The basic means of escape requirements for maisonettes can be similarly enumerated:

1. All maisonettes to have a private entrance hall and stairway where there is no fire risk.
2. All living rooms and bedrooms to have direct access to this hall or stairway.
3. Doors opening on to hall or stairway to be type 3.
4. Hall and stairway to have half-hour fire-resistant walls.
5. In most cases alternative means of escape required, an exception being where a half-hour fire-resistant screen is provided at the head of the stairway.

A typical maisonette example is shown in Fig. IV.21.

Students are encouraged to study the many examples of means of escape planning for flats and maisonettes given at the end of CP3 : Chapter IV: Part 1.

Office Buildings

BSS 5588 : Part 3 covers office buildings of all sizes and heights in the context of means of escape in case of fire and is the mandatory code of practice to be used to comply with Building Regulation B1(2).

The planning procedures for means of escape set out in the code are based on attempting to identify the positions of all possible sources of outbreak of fire and to predict the courses that might follow such an outbreak and in particular the passage of the fire and the passage of the smoke and gases ensuing from the fire. The basic planning objectives are to provide protected escape routes in the horizontal and vertical directions which will enable persons within the building confronted by an outbreak of fire to turn away and make a safe escape without outside assistance.

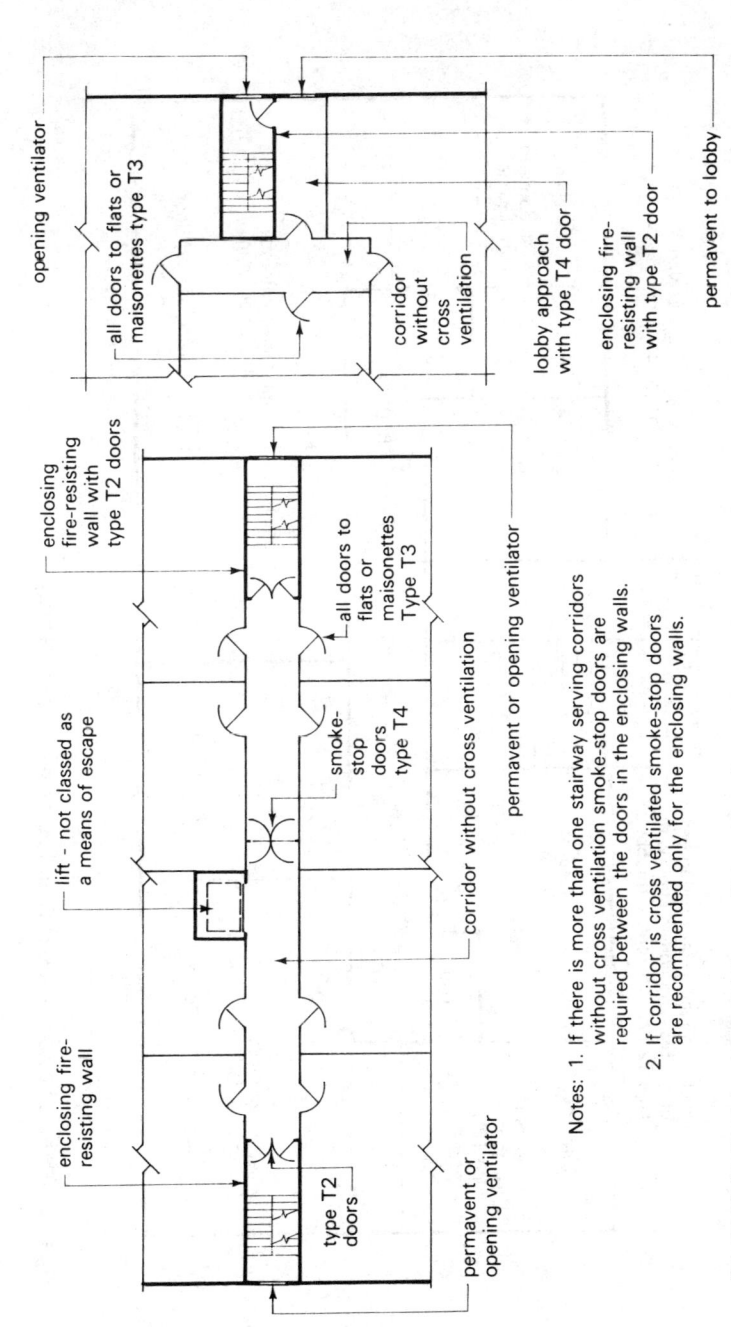

Fig IV.19 Main stairway protection — flats and maisonettes

Notes: 1. If there is more than one stairway serving corridors without cross ventilation smoke-stop doors are required between the doors in the enclosing walls.

2. If corridor is cross ventilated smoke-stop doors are recommended only for the enclosing walls.

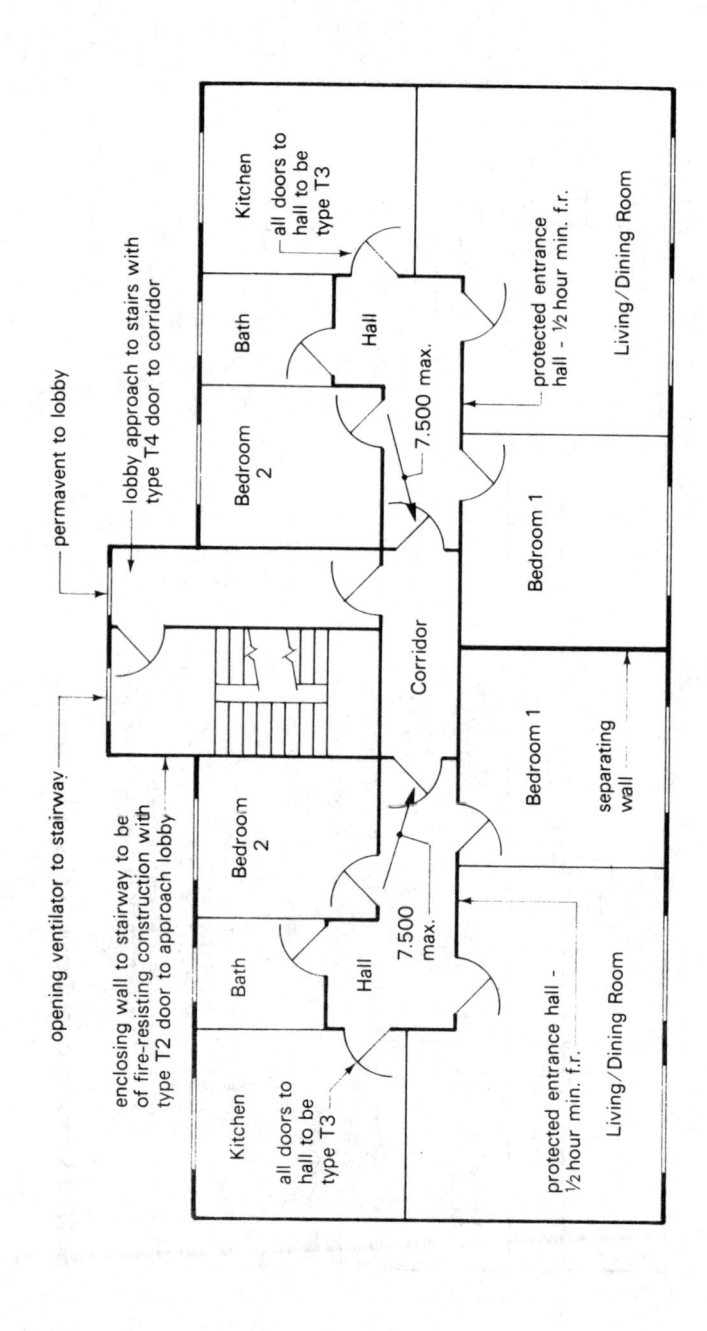

Fig IV.20 Means of escape — example of flats with only one stairway

Lower Floor Plan

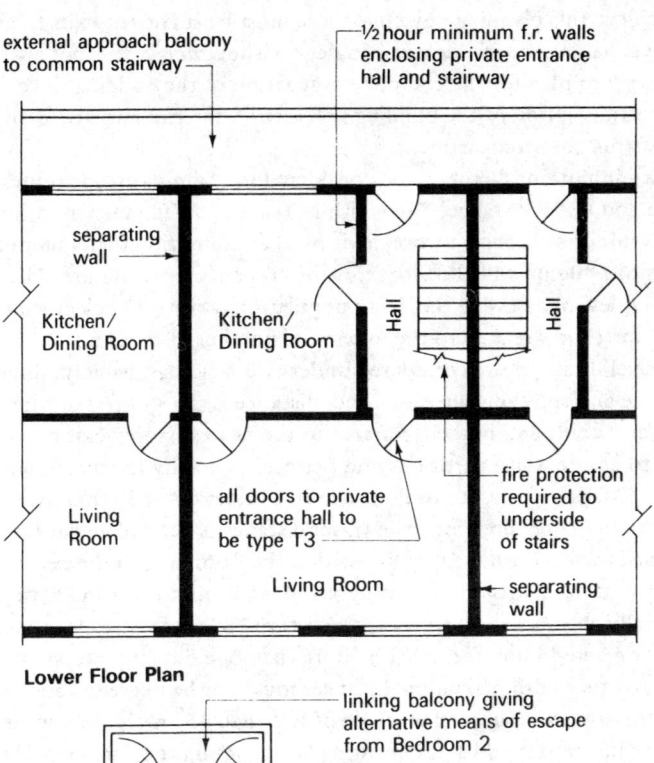

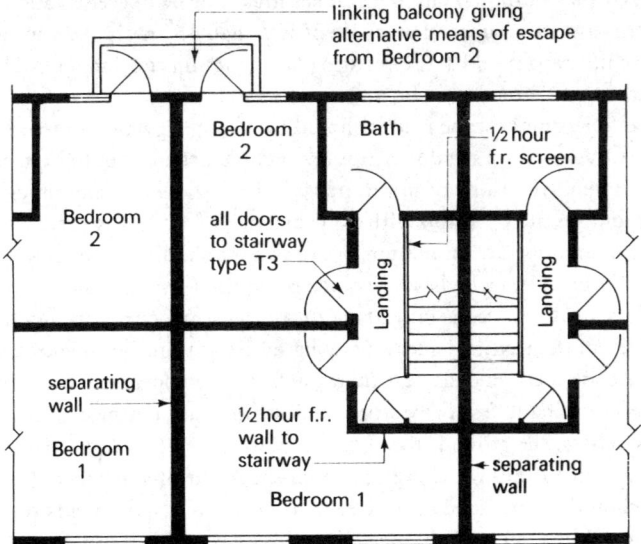

- external approach balcony to common stairway
- ½ hour minimum f.r. walls enclosing private entrance hall and stairway
- separating wall
- Kitchen/Dining Room
- Kitchen/Dining Room
- Hall
- Hall
- Living Room
- all doors to private entrance hall to be type T3
- Living Room
- fire protection required to underside of stairs
- separating wall

Upper Floor Plan

- linking balcony giving alternative means of escape from Bedroom 2
- Bedroom 2
- Bath
- ½ hour f.r. screen
- Bedroom 2
- all doors to stairway type T3
- Landing
- Landing
- separating wall
- Bedroom 1
- ½ hour f.r. wall to stairway
- Bedroom 1
- separating wall

Fig IV.21 Means of escape — maisonette example

To achieve this planning objective there must be sufficient exits to allow the occupants to reach an area of safety without delay. It is not always necessary to plan for the complete evacuation of the building since compartmentation of the building will restrict the fire initially to an area within that compartment.

Two important factors to be considered in planning escape routes are width and travel distance. The width is based upon an evacuation time of 2.5 minutes through a storey exit on the assumption that a unit exit width of 500 mm will allow the flow of 40 persons per minute. The code gives tables for assessing the likely population density and clear widths of escape routes related to the maximum number of persons.

Travel distances are considered under two headings, namely: direct distance and travel distance. A direct distance is the shortest distance from any point within the floor area to the nearest storey exit measured so as to ignore walls, partitions and fittings. Generally the maximum direct distance is 12.000 for escape in one direction or 30.000 for escape in more than one direction. The travel distance is defined as the actual distance travelled from any point within the floor area to the nearest storey exit having regard to the layout of walls, partitions and fittings. Generally the maximum travel distances are 18.000 for escape in one direction and 45.000 for travel in more than one direction. It should be noted that both maximum distances must not be exceeded and that if alternative exits subtend an angle of less than 45° with one another and are not separated by fire-resisting construction they are classified as exits in one direction only.

The horizontal escape routes should be clear gangways with non-slippery even surfaces and if ramps are included these should have an easy gradient of a slope of not more than 1 in 12. The minimum clear headroom should be 2.000 with no projections from walls except for normal handrails. The minimum fire resistance should be not less than half an hour unless a higher resistance is required by the Building Regulations. Limitations on glazing for both escape corridors and doors are given in the text and accompanying tables within the code of practice.

The code recommends that there should not be less than two protected stairways available from any storey unless there are not more than three storeys above the ground storey or the building is of a height of not greater than 11.000 whichever is the lesser. Additional protected stairways should be provided as necessary to meet the requirements for travel distance. Where more than one stairway is provided it is usual to assume that one stairway would be blocked by the fire and therefore the remaining stairways must have sufficient width to cater for the total persons to be evacuated. A typical example of means of escape from an office building is shown in Fig. IV.22.

Shops and Similar Premises

BS 5588 : Part 2 covers all classes of shops and other similar types of premises such as: cafés; restaurants; public houses; and premises where goods are received for treatment, examples being dry cleaners and shoe repairers. The main objective of the code's requirements are to provide safety from fire by means of planning and providing protection of both horizontal and vertical escape routes for any area threatened by fire thus enabling any person confronted by an outbreak of fire to make an unassisted escape. Under the Fire Precautions Act 1971 as amended by the Health and Safety at Work etc., Act 1974 an adequate means of escape is required for all shops and for particular shops a fire certificate will be required.

In planning escape routes there must be sufficient exit facilities, such as protected routes and stairways, to allow the public and staff to reach areas of safety without undue delay. It is not always necessary to plan for complete evacuation immediately an outbreak of fire occurs. Compartmentation of multi-storey shops will limit the rate of fire spread and in such cases a reasonable assumption would be to design for the immediate evacuation of the floor on which the outbreak occurs plus the floor above which is the next portion of the building at risk. Escape widths are based upon an evacuation time of 2.5 minutes and on the assumption that a unit exit width of 500 mm permits the passage of 40 persons per minute. Guidance as to possible population densities is given in the code by means of a table which shows the floor space per person for types of rooms or storeys. Similarly another table gives clear widths for exits related to the total number of persons in a room or storey.

The maximum permitted travel distance is another important factor and can be taken as a maximum distance measured around obstructions such as fixed counters or a direct distance measured over the obstructions. The distances given in a table take into account the number of exits, the relationship of alternative exits and if the floor under consideration is a ground floor or an upper floor. Generally the maximum travel distances in one direction are 18.000 and 45.000 where exits subtend an angle of more than 45°. The maximum direct travel distances being 12.000 and 30.000 for similar conditions.

The vertical escape route via the stairs leading to the final exit leading direct to the open air must also cater for the anticipated number of persons likely to use the stairs for evacuation purposes. A table giving guidance on minimum stair widths related to population density and

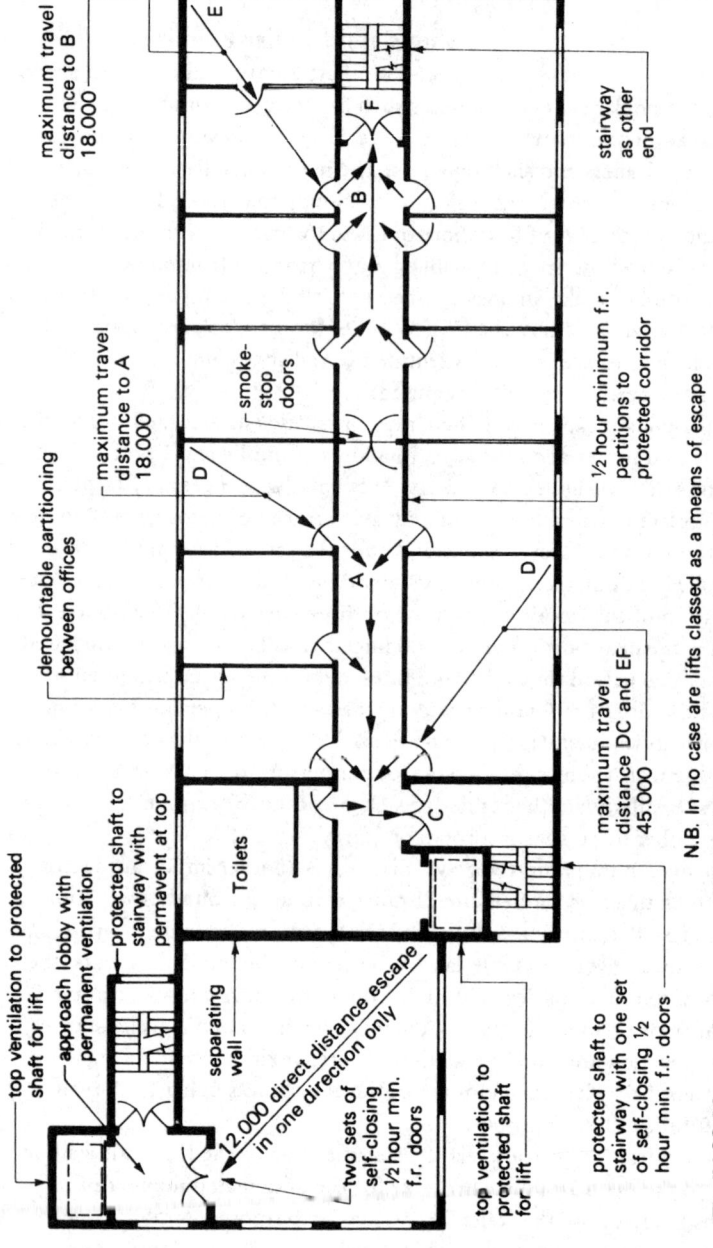

Fig IV.22 Means of escape — office buildings — upper floor example

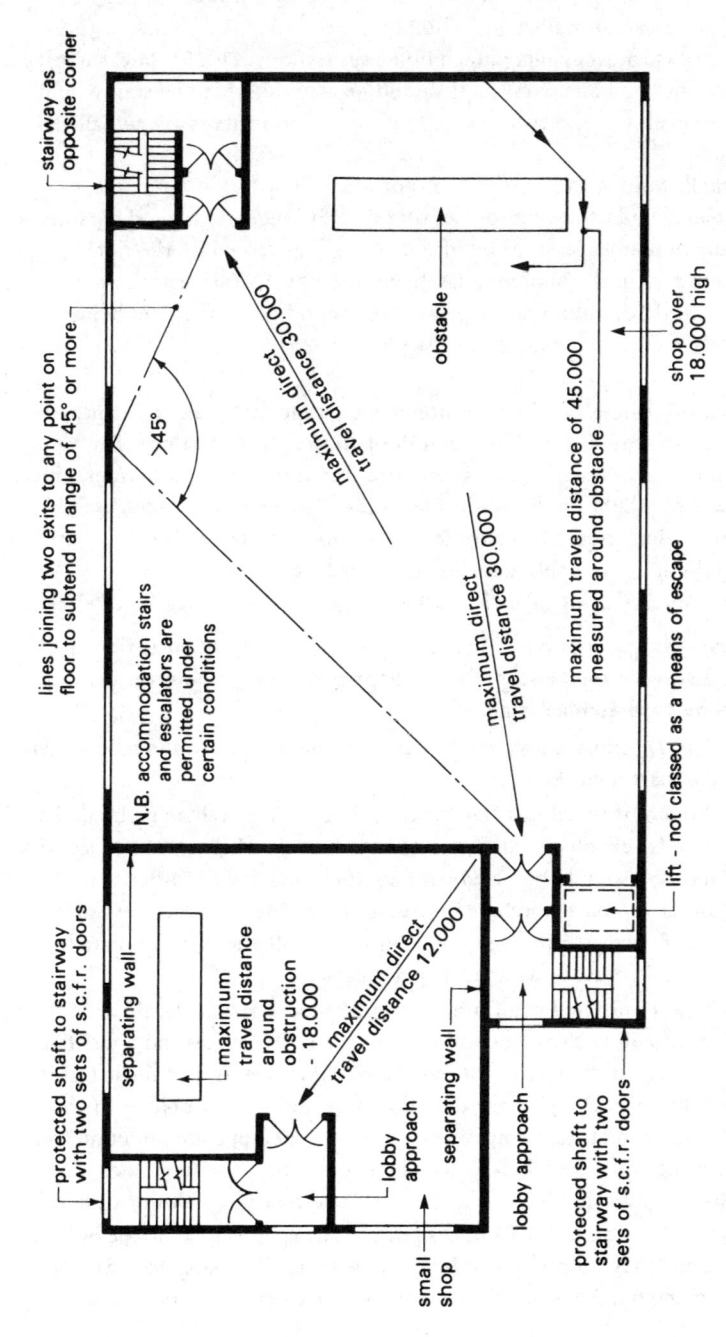

Fig. IV.23 Means of escape — small and large shops — upper floor example

height of building is given in the code. It is recommended that approach lobbies are used for all stairs in buildings over 18.000 high due to the higher risks to occupants in high buildings (see Fig. IV.23). In calculating the number of stairs required it should be assumed that one stair would be inaccessible in the event of a fire if two or more stairs are actually required.

Small shops with floor areas of not more than 280 m^2 in one occupancy and of not more than three storeys high are treated separately. The maximum travel or direct distance being governed by the floor's position within the building. Basement and upper floor distances being 18.000 and 12.000 for single exits with ground-floor distances being 27.000 and 18.000 respectively (see Fig. IV.23).

Staircases: these should be of non-combustible materials and continuous, leading ultimately to the final exit door to the place of safety. The recommended dimensions of going, rise, handrails and maximum number of risers per flight are shown in Fig. IV.24. Fire-resistant glazing is permitted but should be restricted to the portion of the wall above the handrails and preferably designed in accordance with the recommendations of CP 153 : Part 4.

Doors: — in the context of fire, doors are usually classified as fire-check, fire-resisting or smoke-stop. CP3 : Chapter IV : Part 1 however defines doors by type numbers thus:—

Type 1 (T1) must satisfy the Building Regulations as to the requirements for a compartment door.

Type 2 (T2) if fitted in a frame with a 25 mm deep rebate it should have a freedom from collapse and resistance to passage of flame of not less than 30 minutes. It should be fitted with a self-closing device (other than a rising butt) and have unrebated meeting stiles. The door may be single or double leaf, swinging in one or both directions. If there is no rebate to the frame the gap must be as small as practicable.

Type 3 (T3) if fitted in a frame with a 25 mm rebate it should have freedom of collapse for a period not less than 30 minutes and a resistance to the passage of flame for not less than 20 minutes. If of a single leaf it must be hung to swing in one direction only and if of double leaf format to be hung so that the swing of each leaf is in the opposite direction and the meeting stiles are rebated. Doors hung to a frame with rebates of not less than 12 mm should be fitted with an automatic self-closing device.

Type 4 (T4) similiar to T3 doors but the swinging can be in one or both directions. These doors should be fitted with a self-closing device (other than a rising butt) and the frame may be constructed without a rebate.

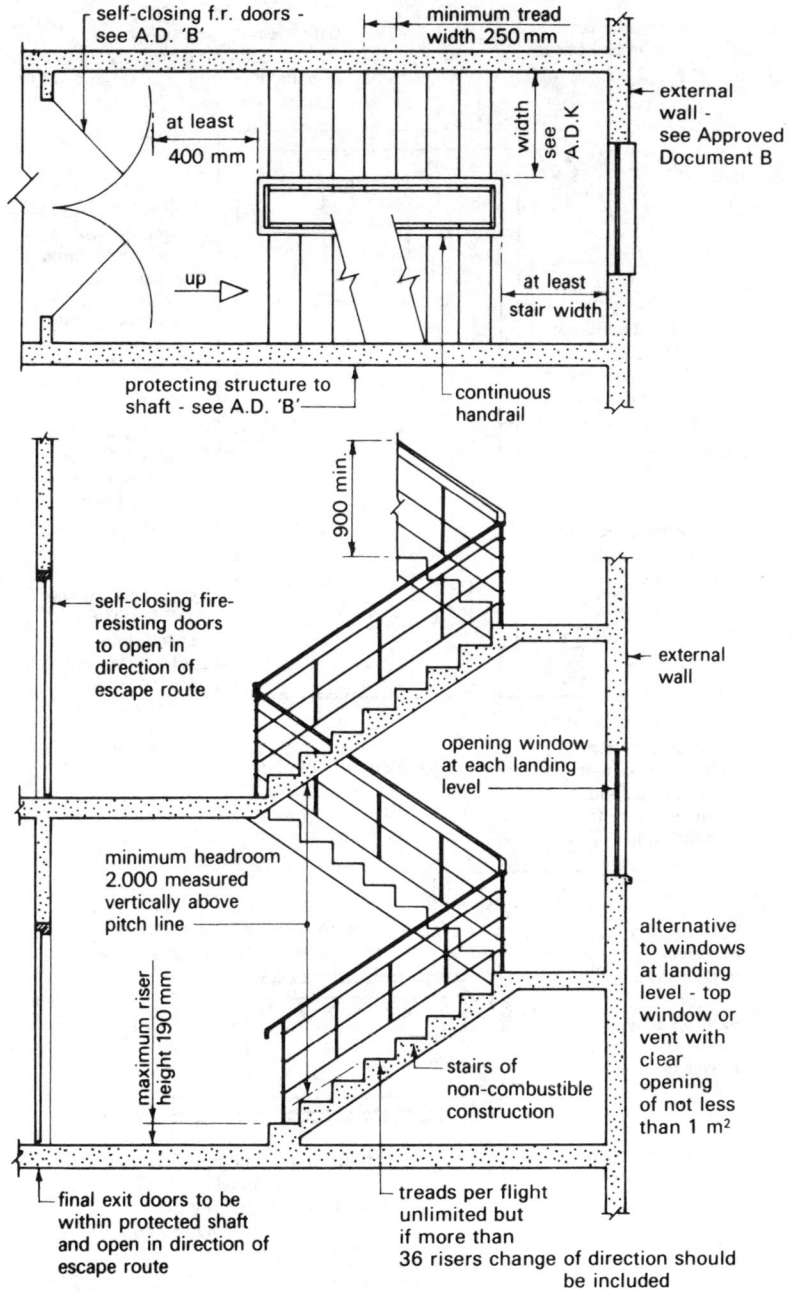

Fig IV.24 Typical escape stair details

- self-closing f.r. doors - see A.D. 'B'
- minimum tread width 250 mm
- at least 400 mm
- width see A.D.K
- up
- at least stair width
- external wall - see Approved Document B
- protecting structure to shaft - see A.D. 'B'
- continuous handrail
- 900 min
- self-closing fire-resisting doors to open in direction of escape route
- external wall
- opening window at each landing level
- minimum headroom 2.000 measured vertically above pitch line
- maximum riser height 190 mm
- stairs of non-combustible construction
- alternative to windows at landing level - top window or vent with clear opening of not less than 1 m²
- final exit doors to be within protected shaft and open in direction of escape route
- treads per flight unlimited but if more than 36 risers change of direction should be included

metal or timber frame

6 mm wired glass panel
½ hour door max. size 1.2 m^2
1 hour door max. size 0.5 m^2

230 min

13 × 13 wood beads encased with a non-combustible cover strip for ½ hour door and non-combustible sub-frame for 1 hour door

3 No. steel hinges to BS 1227 - for 1 hour door broad butts should be used

latch with at least a 12 mm engagement of nib in latch plate

recessed spring type self-closing device

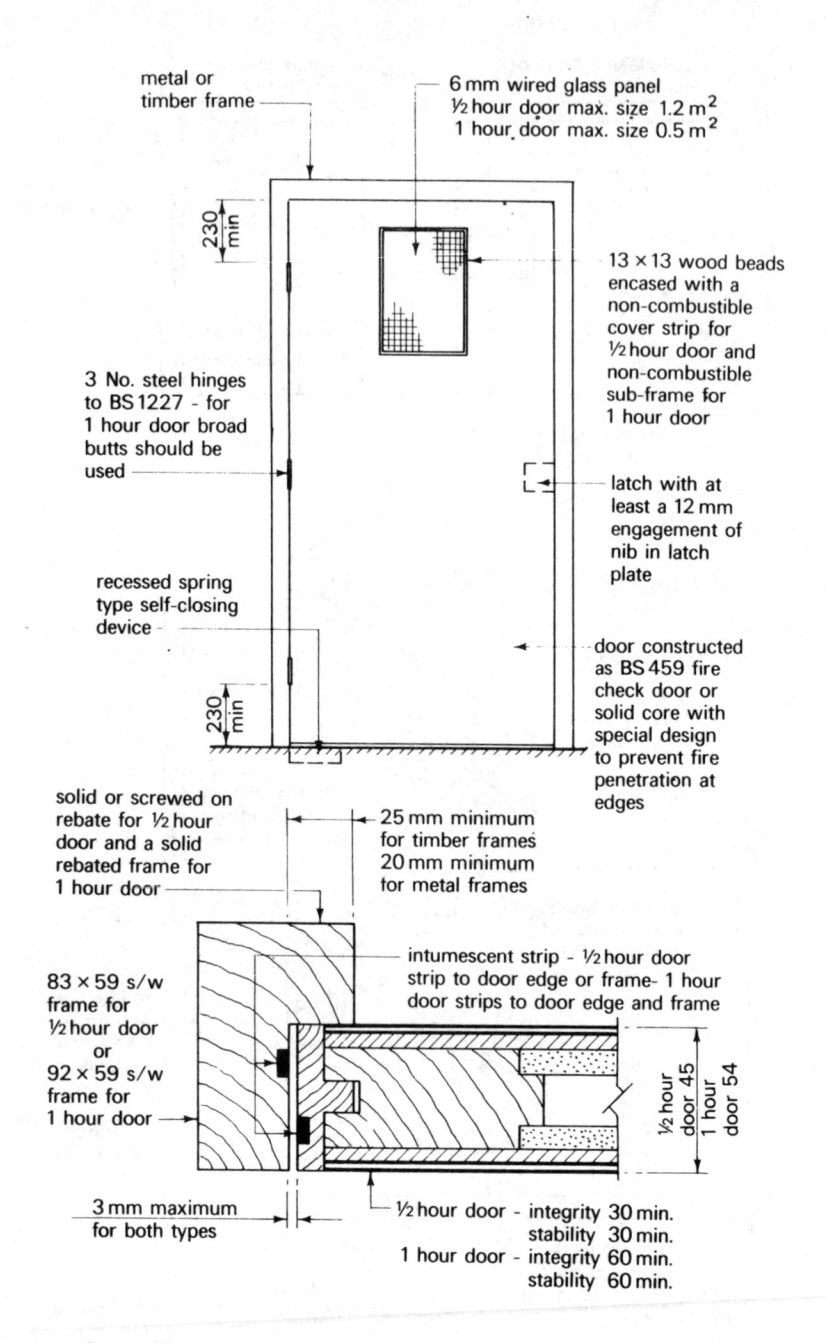

door constructed as BS 459 fire check door or solid core with special design to prevent fire penetration at edges

230 min

solid or screwed on rebate for ½ hour door and a solid rebated frame for 1 hour door

25 mm minimum for timber frames
20 mm minimum for metal frames

83 × 59 s/w frame for ½ hour door or 92 × 59 s/w frame for 1 hour door

intumescent strip - ½ hour door strip to door edge or frame- 1 hour door strips to door edge and frame

½ hour door 45
1 hour door 54

3 mm maximum for both types

½ hour door - integrity 30 min.
stability 30 min.
1 hour door - integrity 60 min.
stability 60 min.

Fig IV.25 Fire-resisting doors

136

Fire-resisting doors fitted in spaces in common use, such as stairways, should be fitted with door closers or spring hinges and hung on hinges complying with the recommendations of BS 1227 with a melting point of metal in excess of 800°C.

Fire-check doors: these are usually made to the recommendations of BS459 : Part 3 (see Fig. III.6, Volume 1) and are similar to fire-resisting doors except they have a lower integrity. Half hour and one hour versions are available which can have 6 mm wired glass panels with a maximum size of 1.2 m^2.

Fire-resisting doors: similar in construction to fire-check doors but greater integrity is achieved by careful design and detail of the gaps between the door edges and the frames by the inclusion of intumescent strips — see Fig.IV.25 for typical details.

Smoke-stop doors: the function of this form of door is obvious from its title and no special requirements are recommended as to door thickness or area of glazing when using 6 mm wired glass. The most vulnerable point for the passage of smoke is around the edges and therefore a maximum gap of 6 mm is usually specified together with a draught excluder seal.

Automatic fire doors: in compartmented industrial buildings it is not always convenient to keep the fire doors in the closed position as recommended. A closed door in such a situation may impede the flow and circulation of men and materials, thus slowing down or interrupting the production process. To seal the openings in the compartment walls in the event of a fire automatic fire doors or shutters can be used. These can be held in the open position under normal circumstances, by counter balance weights or electromagnetic devices and should a fire occur they will close automatically, usually by gravitational forces. The controlling device can be a simple fusible link, a smoke or heat detector which could be linked to the fire alarm system. A typical detail of an automatic sliding fire door is shown in Fig. IV.26.

Pressurised stairways

The traditional methods of keeping escape stairways clear of smoke during a fire by having access lobbies and/or natural ventilations are not always acceptable in modern designs where for example the stairways are situated in the core of the building or where the building has a full air conditioning and ventilation design. Pressurisation is a method devised to prevent smoke

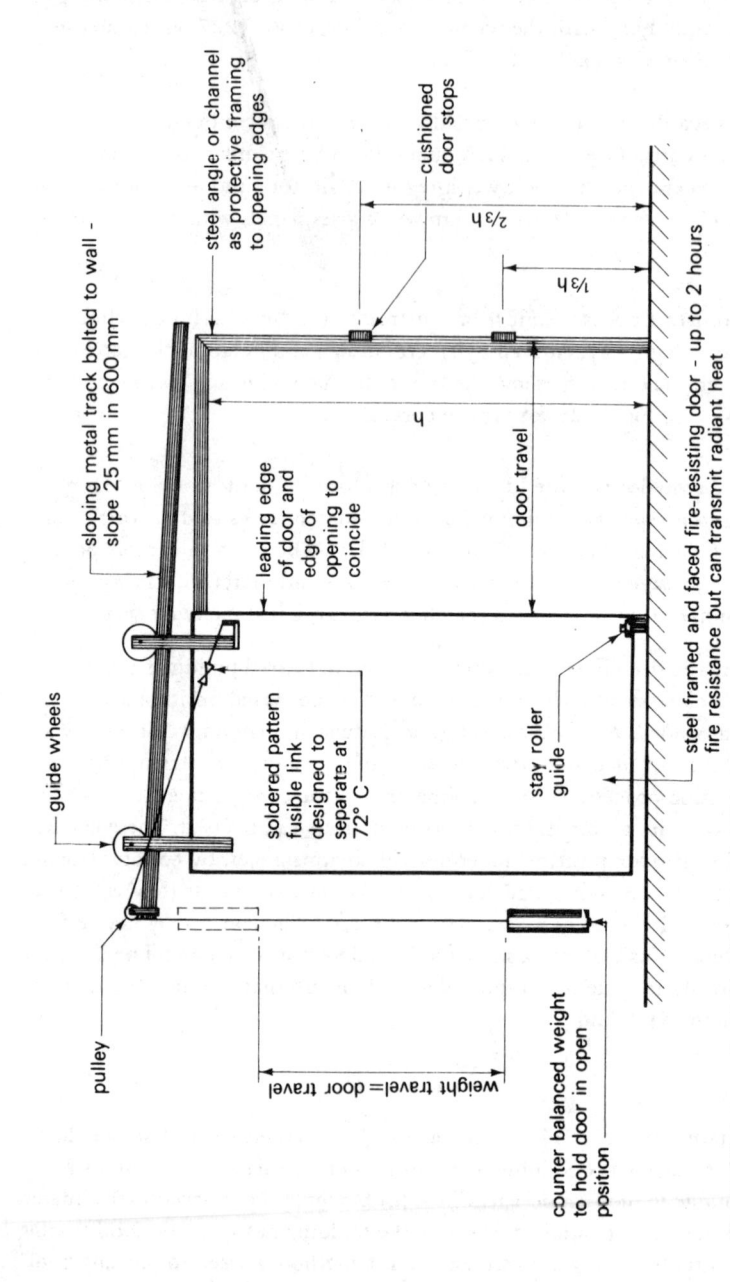

Fig IV.26 Automatic sliding fire door details

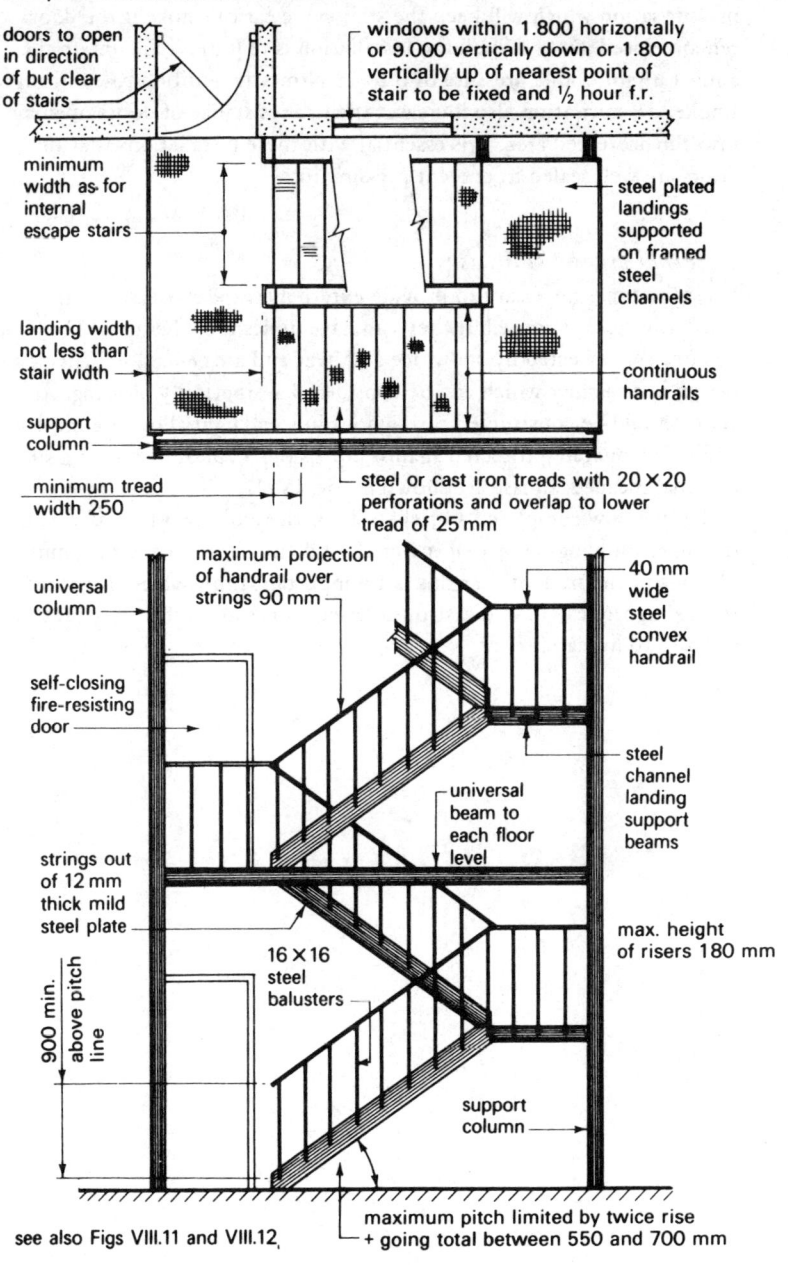

Fig IV.27 Typical external steel escape stair details

logging in an unfenestrated stairway by a system of continuous pressurisation which will keep the stairway clear of smoke if the doors remain closed. Only a small fan installation is required to maintain the almost unnoticeable pressure of 8 to 13 N/m^2. In addition to controlling smoke, pressurisation also increases the fire resistance of doors opening into the protected area. It is essential with these installations that the doors are well sealed to prevent pressure loss.

External escape staircases
It is sometimes necessary to provide external escape staircases as an alternative escape route but they can have undesirable features. They can be adversely affected by snow, ice and heat and are generally difficult to design in a manner which can be considered aesthetically pleasing. All such stairs should be constructed throughout in steel, light alloy or concrete and be adequately protected against corrosion. Typical details of a steel external escape staircase are shown in Fig. IV.27.

Having now completed a brief study of fire and the way it affects the design of buildings choice of materials and circulation patterns it must be obvious to the student that this is an important topic which if ignored or treated lightly can have disastrous consequences not only to a structure but also to human life.

Part V
Claddings to framed structures

13
Cladding panels

Claddings are a form of masking or infilling a structural frame and can be considered under the following headings:

1. Panel walls with or without attached facings — see Chapter 13, Volume 2.
2. Concrete and similar cladding panels.
3. Light infill panels.
4. Curtain walling which can be defined as a sheath cladding which encloses the entire structure and is usually studied in the second year of an advanced course of technology.

All forms of cladding must fulfil the following functions:

1. Be self supporting between the framing members.
2. Provide the necessary resistance to rain penetration.
3. Be capable of resisting both positive and negative wind pressures.
4. Provide the necessary resistance to wind penetration.
5. Give the required degree of thermal insulation.
6. Provide the required degree of sound insulation to suit the building type.
7. Give the required degree of fire resistance.
8. Provide sufficient openings for the admittance of natural daylight and ventilation.
9. Be constructed to a suitable size.

CONCRETE CLADDING PANELS

These are usually made of precast concrete with a textured face in a storey height or undersill panel format. The storey height panel is designed to span vertically from beam to beam and if constucted to a narrow module will give the illusion of a tall building. Undersill panels span horizontally from column to column and are used where a high wall/window ratio is required. Combinations of both formats are also possible.

Concrete cladding panels should be constructed of a dense concrete mix and suitably reinforced with bar reinforcement or steel welded fabric. The reinforcement should provide the necessary tensile resistance to the stresses induced in the final position and for the stresses set up during transportation and hoisting into position. Lifting lugs, positions or holes should be incorporated into the design to ensure that the panels are hoisted in the correct manner so that unwanted stresses are not induced. The usual specification for cover of concrete over reinforcement is 25 mm minimum. If thin panels are being used the use of galvanised or stainless steel reinforcement should be considered to reduce the risk of corrosion.

When designing or selecting a panel the following must be taken into account:

1. Column or beam spacing.
2. Lifting capacities of plant available.
3. Jointing method.
4. Exposure conditions.
5. Any special planning requirements as to finish or texture.

The greatest problem facing the designer and installer of concrete panels is one of jointing to allow for structural and thermal movements and at the same time provide an adequate long term joint — see Chapters 15 and 16. Typical examples of storey height and undersill panels are shown in Figs. V.1 and VI.2.

Where a stone facing is required to a framed structure, possibly to comply with planning requirements, it may be advantageous to use a composite panel. These panels have the strength and reliability of precast concrete panel design and manufacture but the appearance of traditional stonework. This is achieved by casting a concrete backing to a suitably keyed natural or reconstructed stone facing and fixed to the frame by traditional masonry fixing cramps or by conventional fixings — see Fig.V.3.

Thermal insulation can be achieved when using precast concrete panels by creating a cavity as shown in Figs. V.1 and V.2. Alternatively the insulating material can be incorporated in a sandwich cladding panel as shown in Fig. V.10.

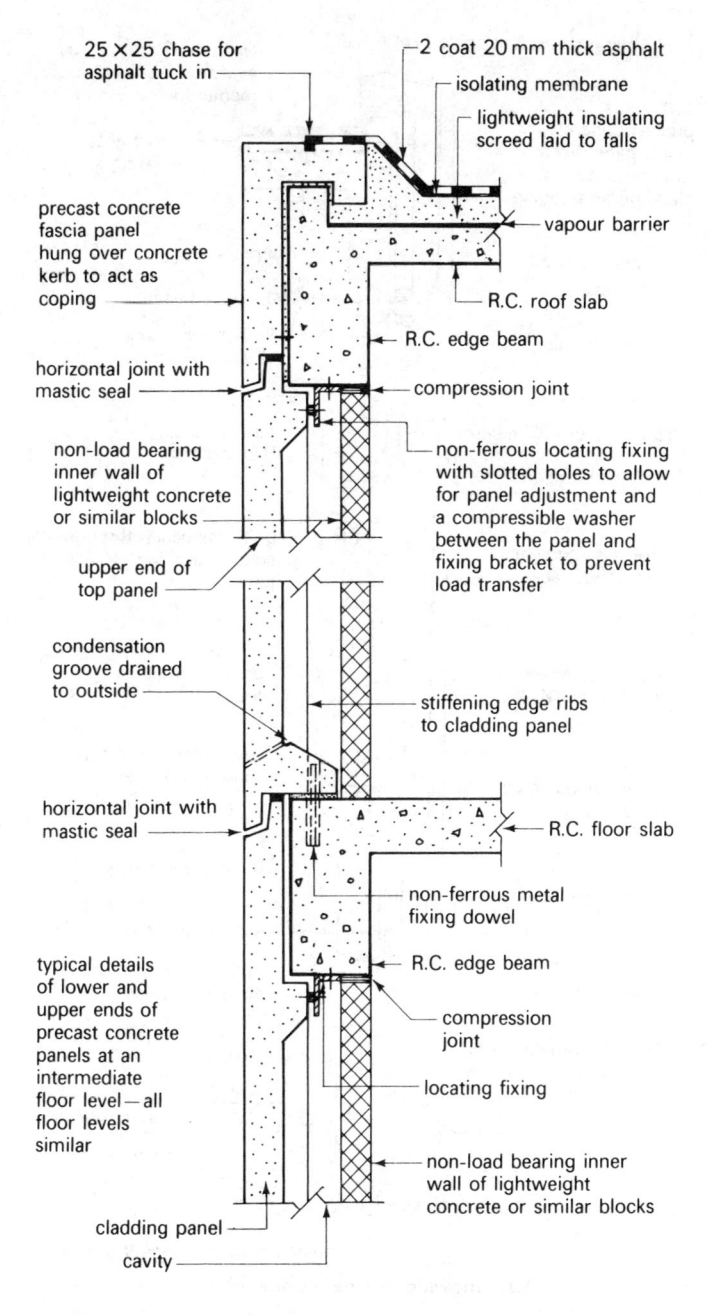

Fig V.1 Typical storey height concrete cladding panel

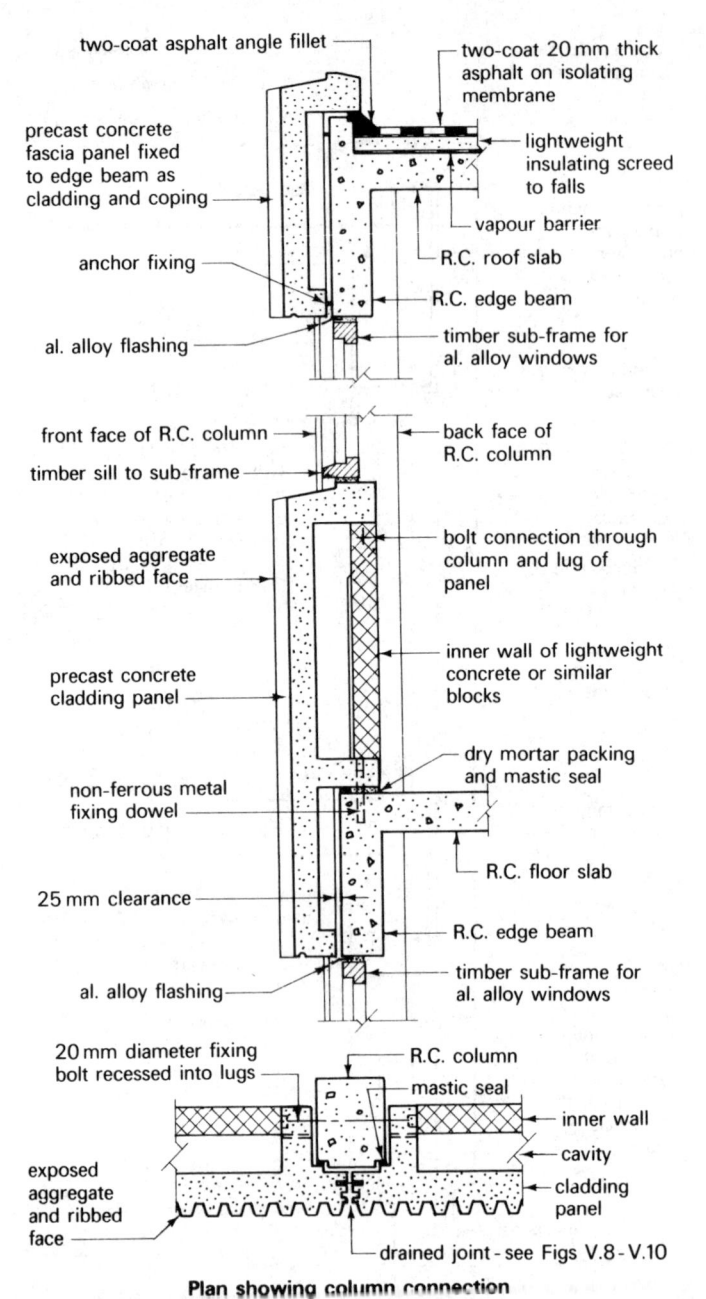

Fig V.2 Typical undersill concrete cladding panel

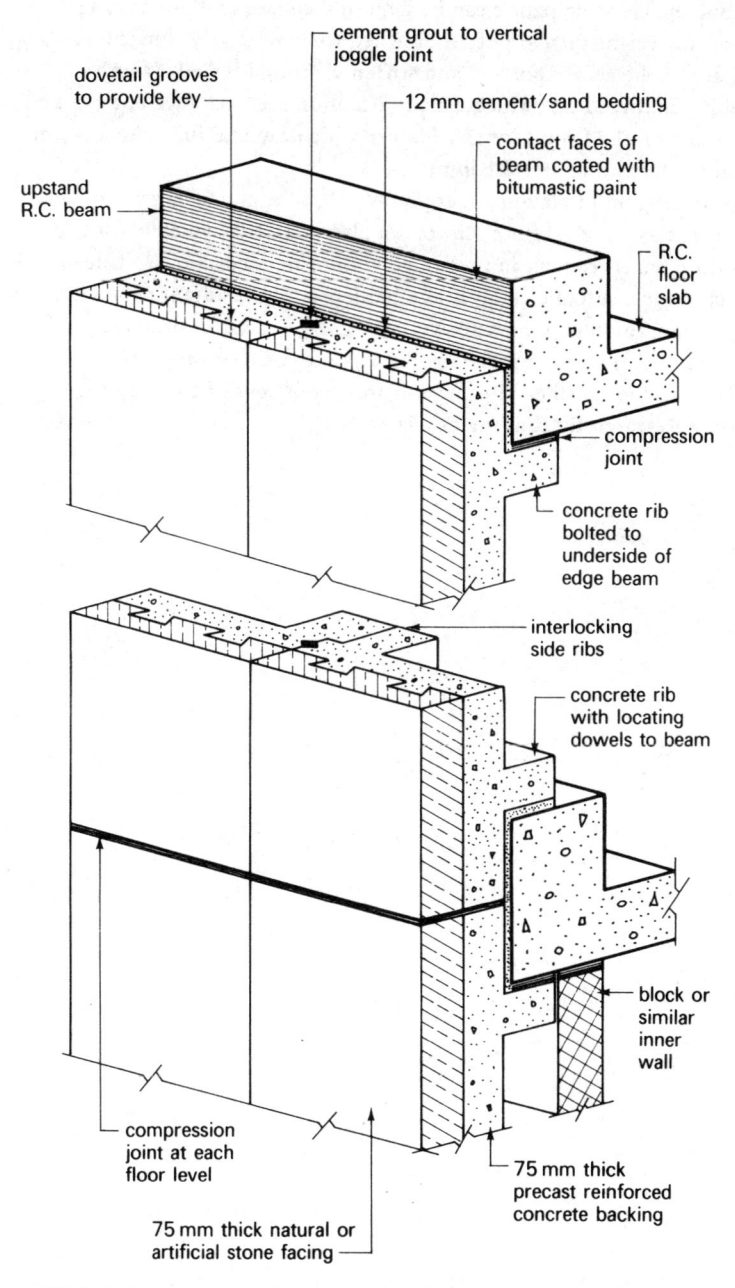

Fig V.3 Typical storey height composite panel

Concrete cladding panels can be large and consequently heavy. To reduce the weight they are often designed to be relatively thin (50 to 75 mm) across the centre portion and stiffened around the edges with suitably reinforced ribs which usually occur on the back face but can be positioned on the front face as a feature which can also limit the amount of water which can enter the joint.

Another form of cladding material which is beginning to gain popularity and acceptance is glass fibre reinforced plastics (GRP) which consists of glass fibre reinforcement impregnated with resin, incorporating fillers, pigments and a suitable catalyst as a hardener. The resultant panels are lightweight, durable, non-corrosive, have good weather resistance, can be moulded to almost any profile and have good aesthetic properties. Students seeking further information are recommended to study the Building Research Establishment Digest 161.

14
Infill panels

The functions of an infill panel are as listed previously for cladding panels in general. Infill panels are lightweight and usually glazed to give good internal natural daylighting conditions. The panel layout can be so arranged to expose some or all of the structural members creating various optical impressions. For example, if horizontal panels are used, leaving only the beams exposed, an illusion of extra length and/or reduced height can be created — see Fig. V.4.

A wide variety of materials or combinations of materials can be employed such as timber, steel, aluminium and plastic. Single and double glazing techniques can be used to achieve the desired sound or thermal insulation. The glazing module should be such that a reasonable thickness of glass can be specified.

The design of the 'solid' panel is of great importance since this panel must provide the necessary resistance to fire, heat loss, sound penetration and interstitial condensation. Most of these panels are of composite or sandwich construction as shown in Figs. V.5 and V.6.

The jointing problem with infill panels occurs mainly at its junction with the structural frame and allowance for moisture or thermal movement is usually achieved by using a suitable mastic or sealant — see Chapter 16.

Most infill panels are supplied as a manufacturer's system, since purpose-made panels can be uneconomic, but whichever method is chosen the design aims remain constant; that is, to provide a panel which fulfils all the required functions and has a low long term maintenance factor. It should be noted that many of the essentially curtain walling systems are

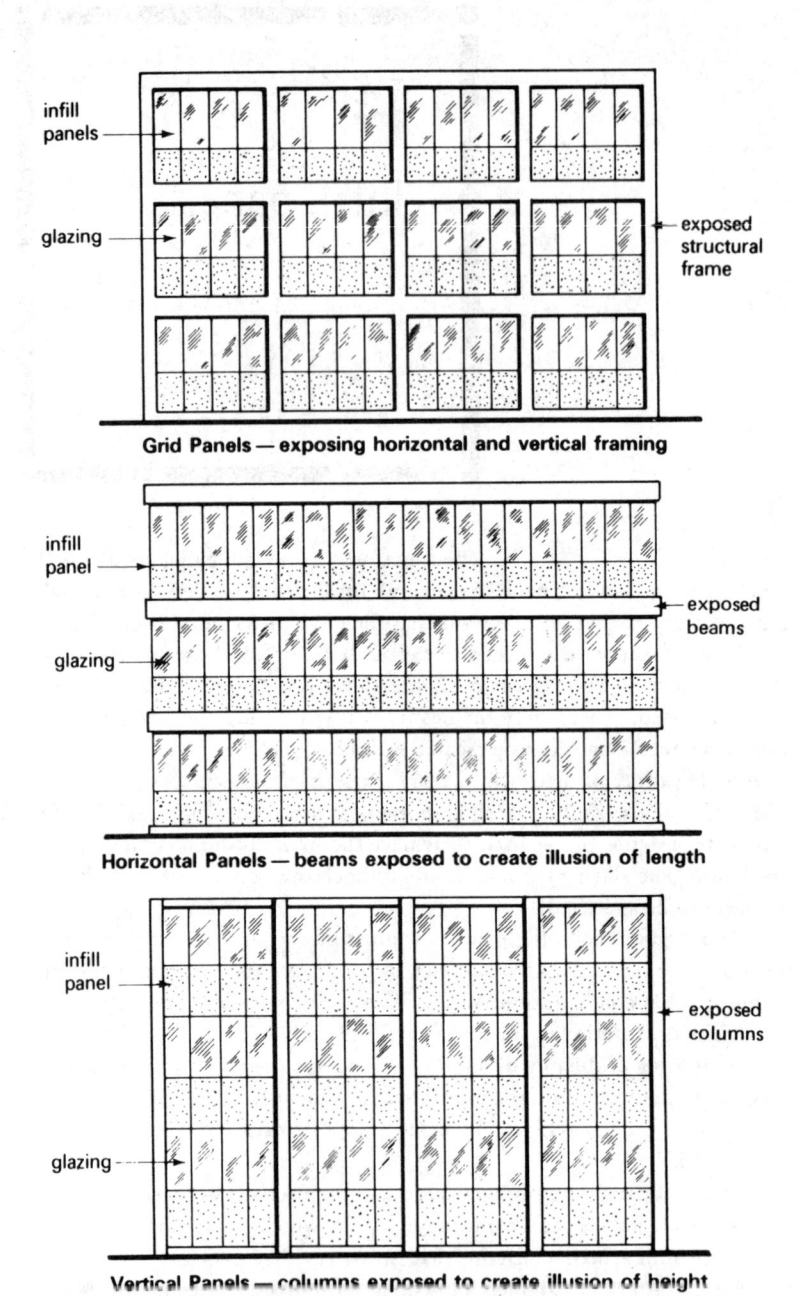

Fig V.4 Typical infill panel arrangements

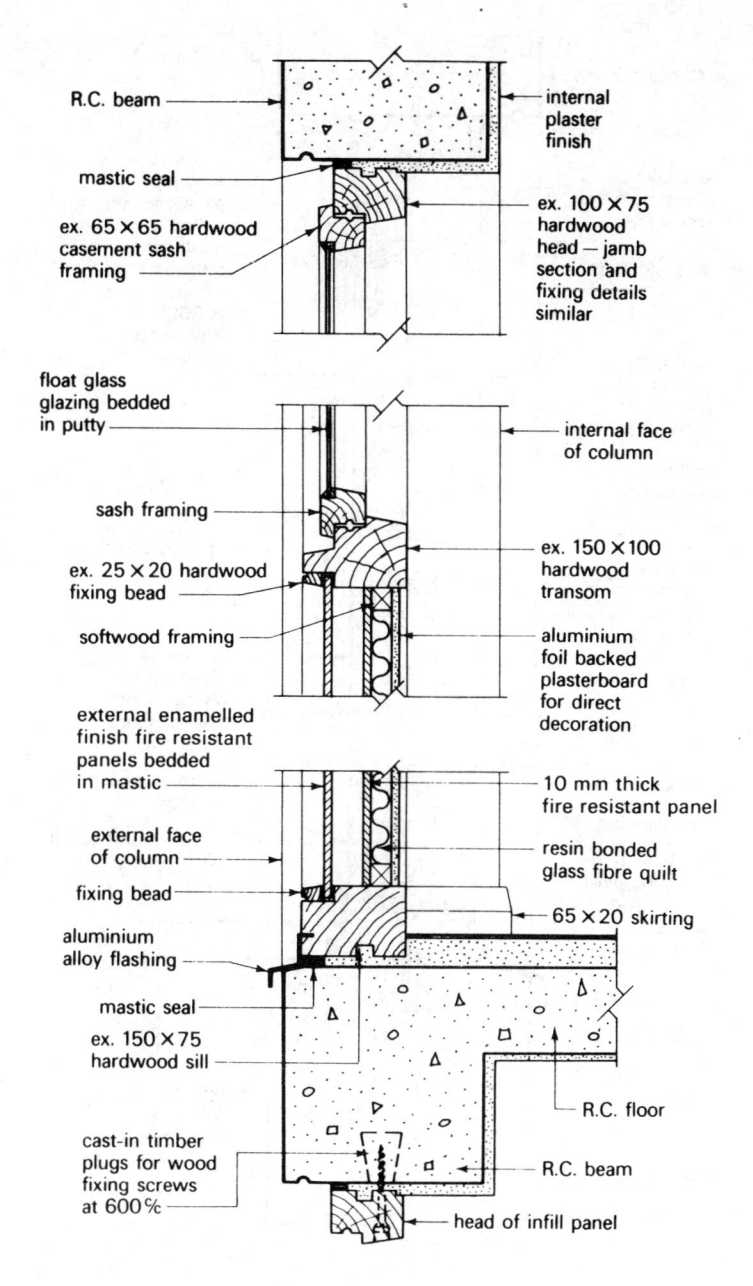

Fig V.5 Typical timber infill panel details

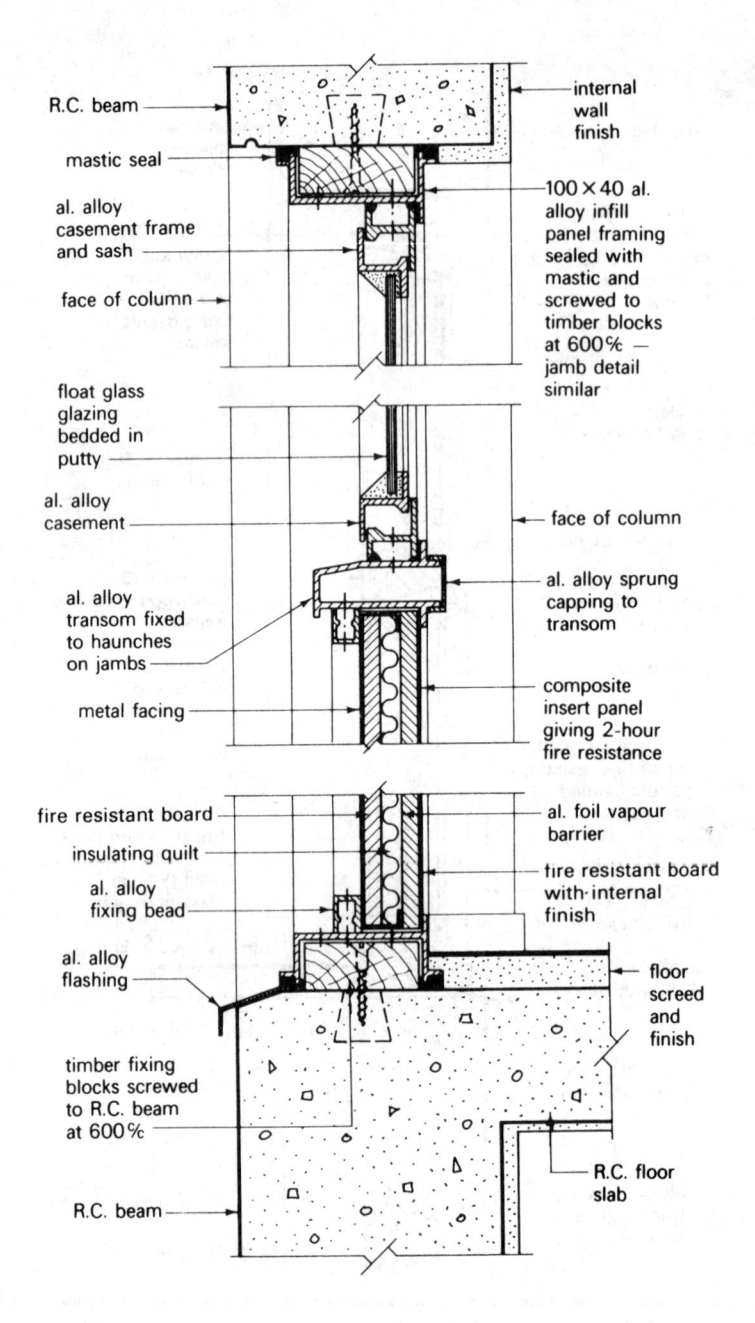

Fig V.6 Typical metal infill panel details

adaptable as infill panels which gives the designer a wide range of systems from which to select the most suitable.

One of the maintenance problems encountered with infill panels and probably to a lesser extent with the concrete claddings is the cleaning of the facade and in particular the glazing. All buildings collect dirt, the effects of which can vary with the material: concrete and masonry tend to accept dirt and weather naturally, whereas impervious materials such as glass do not accept dirt and can corrode or become less efficient.

If glass is allowed to become coated with dirt its visual appearance is less acceptable, its optical performance lessens since clarity of vision is reduced and the useful penetration of natural daylight diminishes. The number of times that cleaning will be necessary depends largely upon the area, ranging from three-monthly intervals in non-industrial areas to six-weekly intervals in areas with a high pollution factor.

Access for cleaning glazed areas can be external or internal. Windows at ground level present no access problems and present only the question of choice of method such as hand cloths or telescopic poles with squeegee heads. Low and medium rise structures can be reached by ladders or a mobile scaffold tower and usually present very few problems. High rise structures need careful consideration. External access to windows is gained by using a cradle suspended from roof level; this can be in the form of a temporary system consisting of counterweighted cantilevered beams from which the cradle is suspended. Permanent systems, which are incorporated as part of the building design, are more efficient and consist of a track on which a mobile trolley is mounted and from which davit arms can be projected beyond the roof edge to support the cradle. A single track fixed in front of the roof edge could also be considered; these are simple and reasonably efficient but the rail is always visible and can therefore mar the building's appearance.

Internal access for cleaning the external glass face can be achieved by using windows such as reversible sashes, horizontal and vertical sliding sashes, but the designer is restricted in his choice to the reach possible by the average person. It cannot be over emphasised that such windows can be a very dangerous hazard unless carefully designed so that all parts of the glazed area can be reached by the person cleaning the windows whilst he remains standing firmly on the floor.

15
Jointing

When incorporating precast concrete cladding panels in a framed structure the problem of making the joints waterproof is of paramount importance. Joints should be designed so that they fulfil the following requirements:

1. Exclude wind, rain and snow.
2. Allow for structural, thermal and moisture movement.
3. Good durability.
4. Easily maintained.
5. Maintain the thermal and sound insulation properties of the surrounding cladding.
6. Easily made or assembled.

Experience has shown that due to bad design, poor workmanship or lack of understanding of the function of a joint has led to water penetration through the joints between cladding panels. To overcome this problem it is essential that both the designer and the site operative fully appreciate the design principles and the need for accurate installation. Suitable joints can be classified under two headings:

1. Filled joints.
2. Drained joints.

Filled joints are generally satisfactory if the cladding panel module is small since if incorporated in large module panels filled joints can crack and allow water to penetrate. This failure is due either to the filling materials being incapable of accommodating movement or a breakdown of adhesion between the filling material and the panel. Research has shown that if the above failures are possible the most effective alternative is the drained joint.

Filled joints: these joints are not easy to construct and rely mainly upon mortars, sealants, mastics or preformed gaskets to provide the barrier against the infiltration of wind and rain. They are limited in their performance by the amount that the sealing material(s) can accommodate movement and to a certain extent their weathering properties such as their resistance to ultra-violet rays. The disadvantages of filled joints can be enumerated thus:

1. Difficulty in making and placing the joints accurately particularly with combinations of materials.
2. Providing for structural, thermal and moisture movement.
3. Only suitable for small module claddings.

For typical detail see Fig. V.7.

Drained joints: these joints have been designed and developed to overcome the disadvantages of the filled joint by:

1. Designing the joint to have a drainage zone.
2. Providing an air-tight seal at the rear of the joint.

Drained joints have two components which must be considered, namely the vertical joint and the horizontal joint.

Vertical joints: consist basically of a deep narrow gap between adjacent panels where the rear of the joint is adequately sealed to prevent the passage of air and moisture. The width of the joint does not significantly affect the amount of water reaching the rear seal since:

1. Most of the water entering the joint (approximately 80%) will do so by following over the face of the panel, the remainder (approximately 20%) will enter the joint directly and most of this water entering the joint will drain down within the first 50 mm of the joint depth. Usually the deciding factor for determining the joint width is the type of mastic or sealant being used and its ability to accommodate movement.
2. Checks on the amount of water entering the drainage zone such as ribs to joint edges, exposed aggregate external surfaces and the use of baffles.

Baffles are loose strips of material such as neoprene, butyl rubber or plasticised PVC which are unaffected by direct sunlight and act as a first line of defence to water penetration. The baffles are inserted, after the panels have been positioned and fixed, either by pulling them through prepared grooves or by direct insertion into the locating grooves from the face or back of panel according to the joint design. Care must be taken when inserting baffles by the pulling method since they invariably stretch during insertion and they must be allowed to return to their original length before trimming off the surplus to ensure adequate cover at the intersection of the vertical and horizontal joint.

The adequate sealing at the back of the joint is of utmost importance since some water will usually penetrate past the open drainage zone or the baffle, and any air movement through the joint seal will also assist the passage of water or moisture. Drained joints which have only a back seal or a baffle and seal can have a cold bridge effect on the internal face giving rise to local condensation: therefore consideration must be given to maintaining the continuity of the thermal insulation value of the cladding — see typical details in Figs. V.7 and V.8.

Horizontal joints: these are usually in the form of a rebated lap joint, the upper panel being lapped over the top edge of the lower panel. As with the vertical joints, the provision of an adequate back seal to prevent air movement through the joint is of paramount importance. The seal must also perform the function of a compression joint, therefore the sealing strip is of a compressible material such as bituminised foamed polyurethane or a preformed cellular rubber strip.

The profile of the joint is such that any water entering the gap by flowing from the panel face or by being blown in by the wind is encouraged to drain back on to the face of the lower panel. The depth of joint overlap is usually determined by the degree of exposure and ranges from 50 mm for normal exposure to 100 mm for severe exposure — see Fig. V.9 for typical details. It must be noted that the effective overlap of a horizontal joint is measured from the bottom edge of the baffle in the vertical joint to the seal and not from the rebated edge of the lower panel — see Fig. V.10.

Intersection of joints: this is an important feature of drained joint design and detail since it is necessary to shed any water draining down the vertical joint on to the face of the lower panels where the vertical and horizontal joints intersect because the joints are designed to cater only for the entry of water from any one panel connection at a time. The usual method is to use a flashing starting at the back of the panel, dressed over and stuck to the upper edge of the lower panel as shown in Fig. V.10. The choice of material for the flashing must be carefully considered since it must accept the load of the upper panel and any movements made whilst the panel is

Filled joint

- precast concrete cladding
- mastic or sealant
- backing of compressible plastic material
- external face gap filling material

Drained joint with water bar

- mastic seal to sprung al. alloy water bar
- vertical dpc
- insulation to prevent cold bridge
- internal wall finish
- drainage zone
- precast concrete cladding

Drained joint with preformed seal

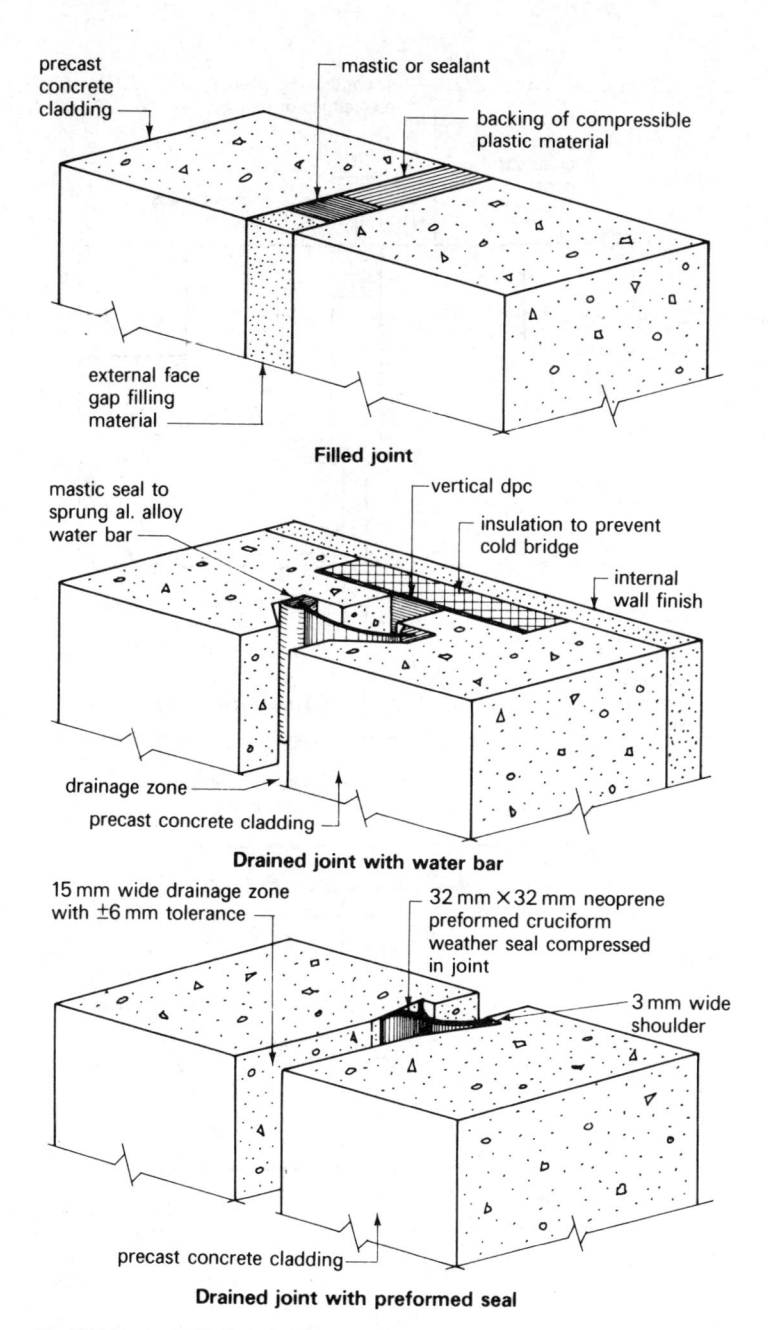

- 15 mm wide drainage zone with ±6 mm tolerance
- 32 mm × 32 mm neoprene preformed cruciform weather seal compressed in joint
- 3 mm wide shoulder
- precast concrete cladding

Fig V.7 Typical filled and drained joints

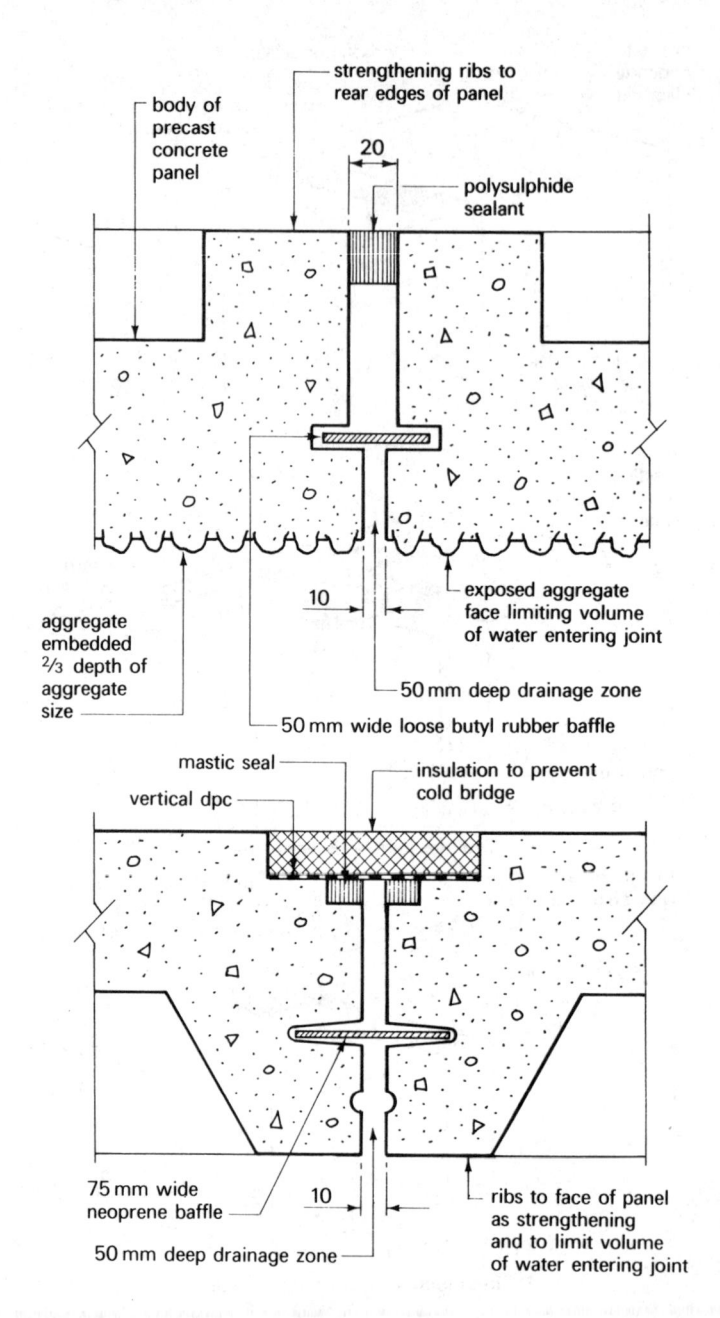

Fig V.8 Typical drained joints using baffles

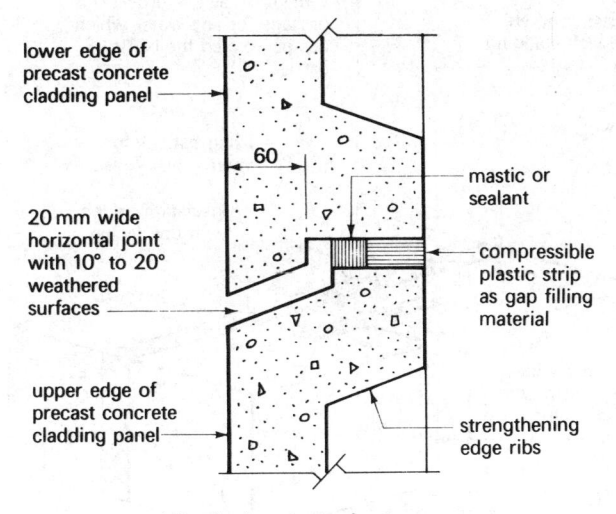

Joint for low to moderate exposure

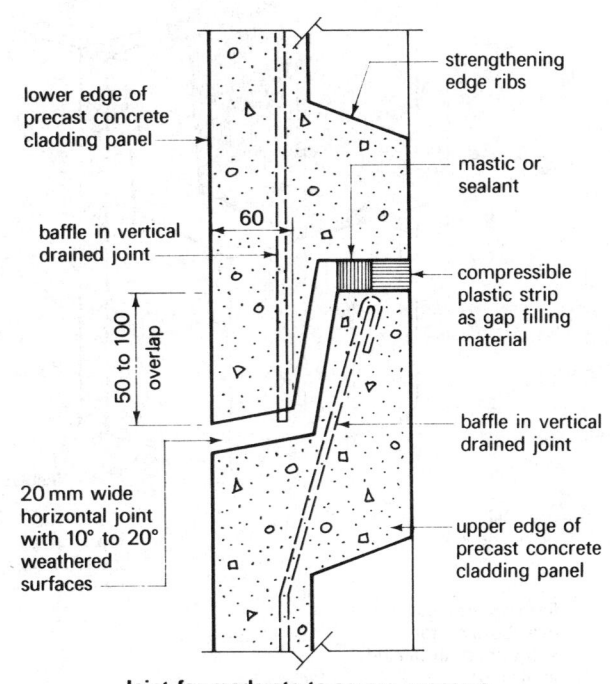

Joint for moderate to severe exposure

Fig V.9 Typical horizontal joints

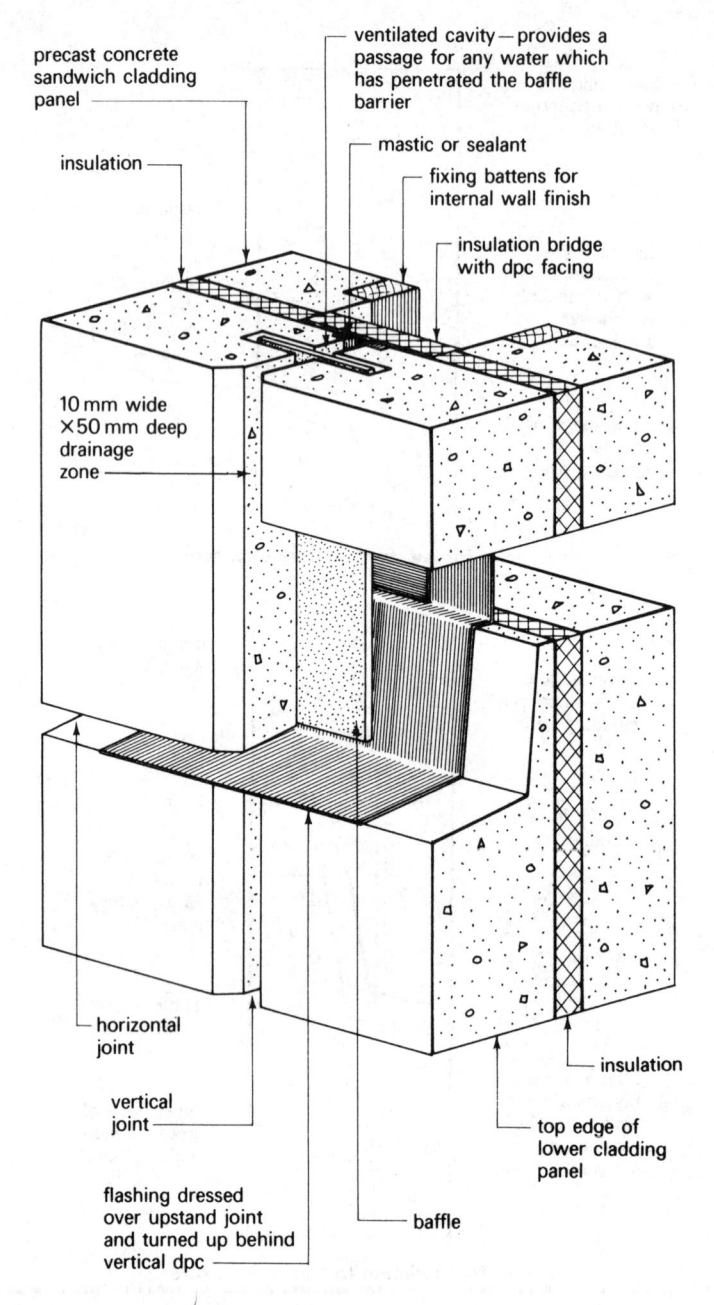

Fig V.10 Typical drained joint intersection detail

positioned and secured. Also it should be a material which is durable but will not give rise to staining of the panel surface. Experience has shown that suitable materials are bitumen coated woven glasscloth and synthetic rubber sheet.

16
Mastics and sealants

Materials which are to be used for sealing joints whether in the context of claddings or for sealing the gap between a simple frame and the reveals has to fulfil the following requirements:

1. Provide a weathertight seal.
2. Accommodate movement due to thermal expansion, wind loadings, structural movement and/or moisture movement.
3. Accommodate and mask tolerance variations.
4. Stable; that is, to remain in position without slumping.
5. Should not give rise to the staining of adjacent materials.

Mastics and sealants have a limited life ranging from 10 to 25 years, which in most cases is less than the design life of the structure, therefore all joints must be designed in such a way that the seals can be renewed with reasonable ease, efficiency and cost. The common form of sealing material, called putty, is unsuitable for many applications since it hardens soon after application and cannot therefore accept the movements which can be expected in claddings and similar situations. It was this inability to accommodate movement that led to the development of mastics and sealants.

Mastics: materials which are applied in a plastic state and form a surface skin over the core which remains pliable for a number of years.

Sealants: capable of accommodating greater movement than mastics, are more durable but dearer in both material and installation costs. They are applied in a plastic state and are converted by chemical reactions into an elastomer or synthetic rubber.

A wide variety of mastics and sealants are available to the designer and builder giving a variety of properties and applications. A text of this nature does not lend itself to a complete analysis of all the types available but only to a comparison of the most common grades.

Butyl mastics: a basic mastic with the addition of butyl rubber or related polymers making it suitable for glazing. They have a durability of up to 10 years and can accommodate both negative and positive movements up to 5% using a maximum joint width of 20 mm and a minimum joint depth of 6–10 mm.

Oil-bound mastics: the most widely used mastic, made with non-drying oils and has similar properties to the butyl mastics but in some cases a better durability of up to 15 years. Joint width 25 mm maximum although some special grades can be up to 50 mm, but in all cases the minimum joint depth is 12 mm.

Two-part polysulphide sealant: the best known elastomer sealant having an excellent durability of 25 years or more with a movement accommodation of approximately 15%. It is supplied as two components, a polysulphide base and a curing agent, which are mixed on site shortly before use, the mixture having a pot life of approximately 4 hours. The two parts will chemically react to form a synthetic rubber and should be used with a maximum joint width of 25 mm and a minimum joint depth of 6 mm when used in conjunction with metal or glass and a 10 mm minimum joint depth when used in connection with concrete. Two part polysulphide sealants should conform to the recommendations of BS 4254.

One-part polysulphide sealant: does not require premixing on site before application and is converted into a synthetic rubber by absorbing moisture from the atmosphere. The joint sizing is similar to that described for a two-part polysulphide but the final movement accommodation is less at approximately plus or minus 12½%. The curing process is slow, 1–2 months, and during this period the movement accommodation is very low. One-part polysulphides are generally specified for pointing and where early or excessive movements are unlikely to occur.

Silicone rubber sealant: a one-part sealing compound which converts to an elastomer by absorbing moisture from the atmosphere and has similar properties to the one-part polysulphide. It is available as a pure white or as a translucent material which makes it suitable for the sealing of internal tiling.

Mastics and sealants can be applied in a variety of ways such as gun or knife applications. The most common method is by hand-held gun using disposable cartridges fitted into the body of the applicator. Most guns can be fitted with various nozzles to produce a neat bead of the required shape and size. Mastics are also available for knife application, the material being supplied in tins or kegs up to 25 kg capacity.

The joint should be carefully prepared to receive the mastic or sealant by ensuring that the contact surfaces are free from all dirt, grease and oil. The contact surfaces must be perfectly dry and in some instances they may have to be primed before applying the jointing compound.

It should be noted that as an alternative to using mastics and sealants for jointing, pre-formed gaskets of various designs and shapes are available — see Fig. V.7. These can have better durability than mastics and sealants but the design and manufacture of the joint profile requires a high degree of accuracy if a successful joint is to be obtained.

Part VI
Factory buildings

17
Roofs

The general concepts of roof functions, construction techniques and methods of covering have already been covered in the first two years of study (see relevant sections in volumes 1 and 2). The study in a typical third year course is concerned mainly with certain building types such as small to large industrial or factory buildings. Buildings of this nature set the designer two main problems:

1. Production layout at floor level and the consequent need for large unobstructed areas by omitting as far as practicable internal roof supports such as load bearing walls and columns.
2. Provision of natural daylight from the roof over the floor area below.

The amount of useful daylight which can penetrate into a building from openings in side and end walls is very limited and depends upon such factors as height of windows above the working plane or surface, sizes of windows and the arrangement of windows. Buildings with spans in excess of 18.000 m will generally need some form of overhead supplementary lighting. In single storey buildings this can take the form of glazed rooflights but in multi-storey buildings the floors beneath the top floor will have to have the natural daylighting from the side and end walls supplemented by permanent artificial lighting.

The factors to be considered when designing or choosing a roof type or profile for a factory building in terms of rooflighting are:

1. Amount of daylight required.
2. Spread of daylight required over the working plane.

3. Elimination of solar heat gain and solar glare.

The amount of daylight required within a building is usually based upon the daylight factor, which can be defined as 'the illumination at a specific point indoors expressed as a percentage of the simultaneous horizontal illumination outdoors under an unobstructed overcast sky'. The minimum daylight factor for factories recommended by the 'Illuminating Engineering Society' is 5%. The designer can calculate the daylight factor achieved by a particular roof profile and glass area by using a BRS daylight protractor (see BRE Digests 41 and 42) but an approximation can be made by using a rule of thumb which gives a ratio of one-fifth glass to floor area thus:

assume a daylight factor of 6% is required over a floor area of 500 m^2
then area of glass required
  = daylight factor × floor area × 5
  = $\frac{6}{100}$ × 500 × 5
  = 150 m^2

It must be emphasised that this one-fifth rule of thumb is a preliminary design aid and the result obtained should be checked by a more precise method.

For an even spread of light over the working plane the ratio between the spacing and height of the rooflights is particularly important unless a monitor roof is used which gives a reasonable even spread of light by virtue of its shape. For typical ratios see Fig. VI.2. If the ratios shown in Fig. VI.2 are not adopted the result could give a marked difference in illumination values at the working plane, resulting in light and darker areas.

Another factor to be considered when planning rooflighting is the amount of obstruction to natural daylighting which could be caused by services and equipment housed in the roof void. Also the gradual deterioration of the effectiveness of the glazed areas due to the collection of dirt on the surfaces.

The elimination of solar heat gain and solar glare can be achieved in a number of ways such as fitting reflective louvres to the glazed areas and treating the surface with a special thin paint wash to act as a diffusion agent. Probably the best solution is to choose a roof profile such as the northlight roof which is orientated away from the southern aspect. It must be appreciated that due to other factors such as site size, site position and/or planning requirements the last solution is not always possible.

NORTHLIGHT ROOFS

This form of roof profile is asymmetric; the north orientated face is pitched at an angle of between 50° and 90° and is covered with glass, usually in the form of patent glazing. The south orientated face of the roof is pitched at an angle of between 20° and 30° and is covered with profiled asbestos cement sheeting or similar lightweight covering attached to purlins. The structural roof members can be of timber, steel or precast concrete formed as a plane frame in single or multi-span format and spaced at 4.500 to 6.000 m centres according to the spanning properties of the purlins — see Fig. IV.1.

Single span northlight roofs in excess of 12.000 m span are generally unacceptable since the void formed by the roof triangulation is very large and since it is not divorced from the main building it has to be included in the total volume to be heated. The solution is to use a series of smaller plane frames to form a multiple northlight roof which will reduce the total volume of the roof considerably. It should be noted that there is practically no difference in the total roof area to be covered whichever method is adopted; however, multiple roofs will require more fittings in the form of ridge pieces, eaves closures and gutters.

Using a system of multiple northlight frames a valley is formed between each plane frame and this must be designed to collect and discharge the surface water run off from both sides of the valley — see Fig. VI.3. Internal support for the ends of adjacent frames can be obtained by using a valley beam spanning over internal columns, the spacing of these supporting columns being determined by the spanning properties of the chosen beam. If the number of internal columns required becomes unacceptable in terms of floor layout and circulation an alternative arrangement can be used consisting of a lattice girder housed between the apex and bottom tie of the roof frame — see Fig. VI.1. This method will enable economic spans of up to 30.000 m to be achieved. The inclusion of a lattice girder in this position will create a cantilever northlight roof truss and if the same principle is adopted for a multispan symmetrically pitched roof truss it is usually called an umbrella roof. The only real disadvantage of this alternative method is a slight increase in shadow casting caused by the lattice beam member and a small increase in long term maintenance such as painting.

MONITOR ROOFS

This form of roof is basically a flat roof with raised portions glazed on two faces which are called monitor lights. The roof covering is usually some form of lightweight metal decking covered with asphalt or built-up roofing felt. The glazed areas, like those in a

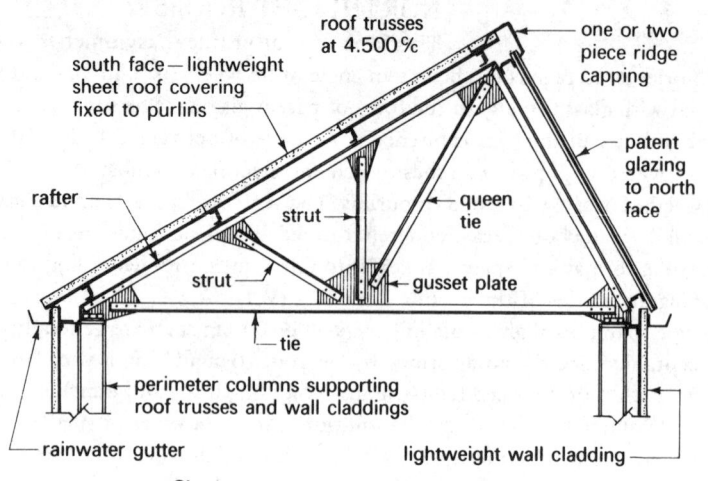

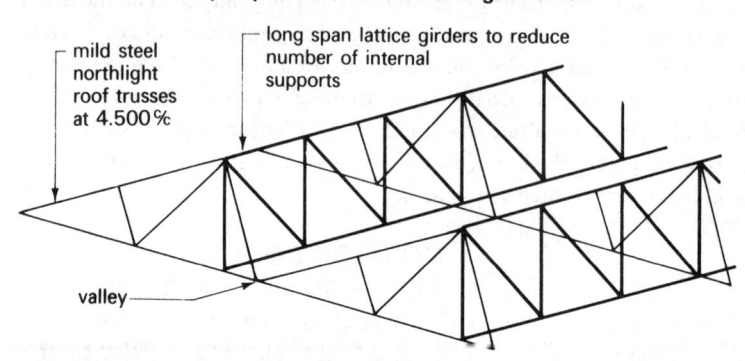

Fig VI.1 Typical northlight roof profiles

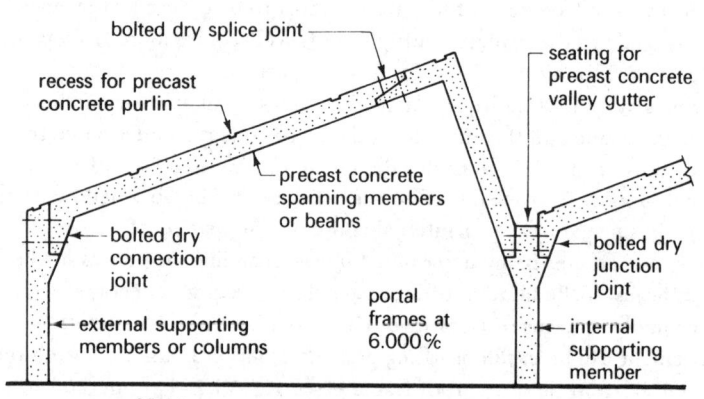

Northlight roofs

even spread of light ratio
$S:H \ngtr 2:1$

Symmetrically pitched roofs

even spread of light ratios
$S_1:H_1 \ngtr 2:1 \quad S_2:H_2 \ngtr 2.5:1 \quad S_3:H_3 \ngtr 1:1$

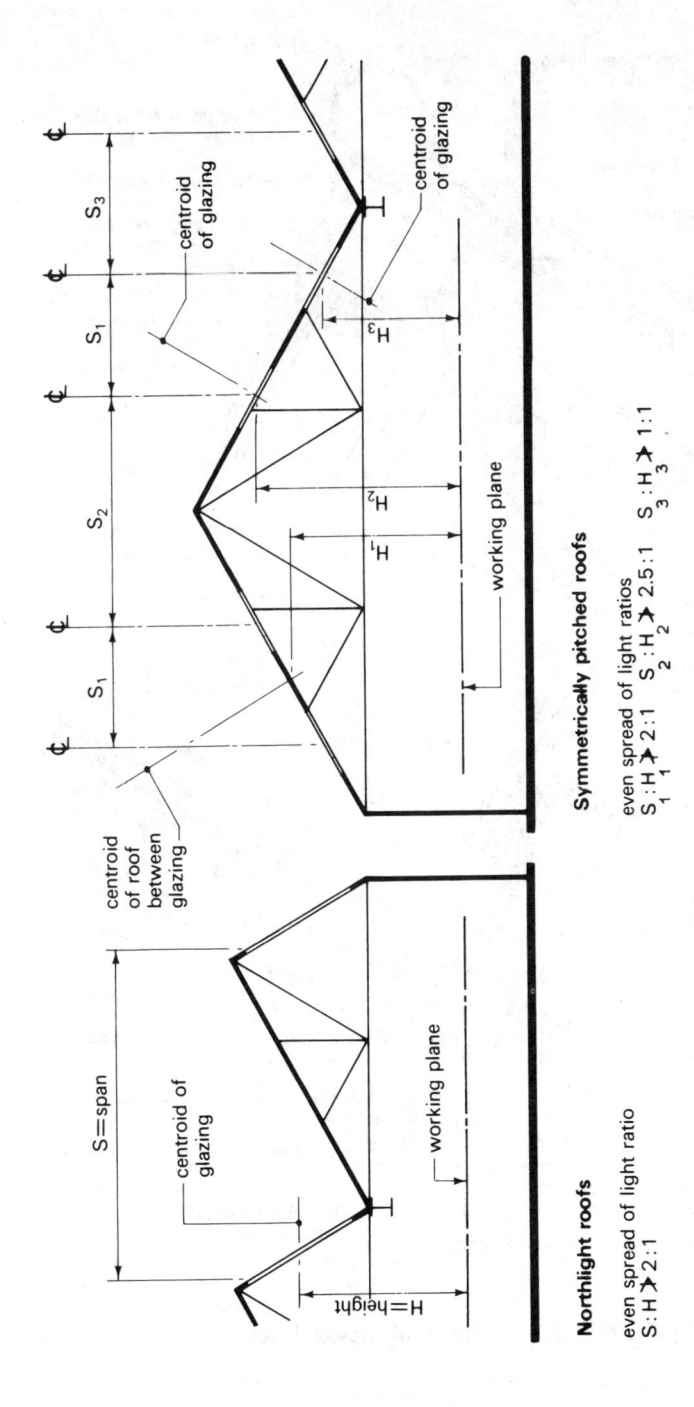

Fig VI.2 Roof profiles - even spread of light ratios

Ridge detail

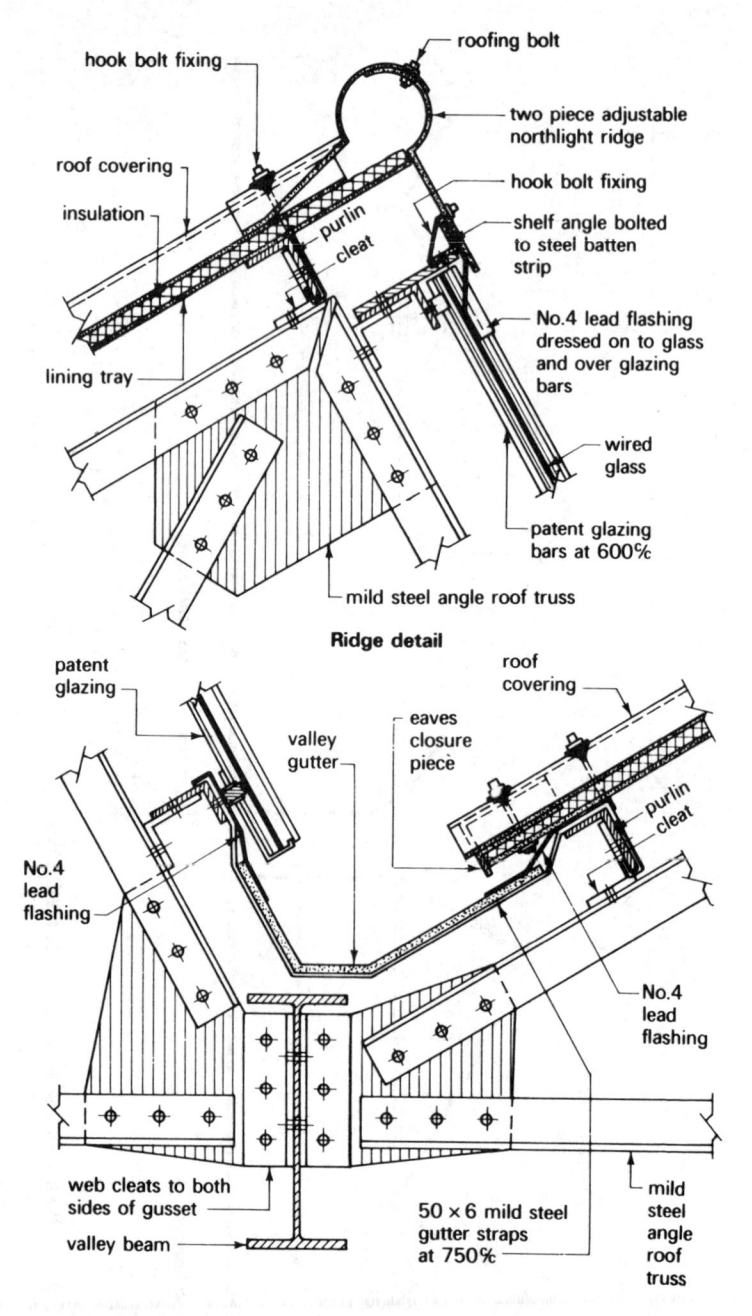

Fig VI.3 Multiple northlight roof - typical details

northlight roof, are usually of patent glazing and are generally pitched at an angle of between 70° and 90°.

Monitor lights give a uniform distribution of natural daylight with a daylight factor of between 5% and 8%. Their near vertical pitch does not give rise to solar glare problems and therefore orientation is not of major importance. The void is considerably less than either the symmetrically pitched or northlight roof which gives a more economic solution to heating design problems. The flat ceiling areas below the monitor lights will give a better distribution of artificial lighting than pitched roofs, also the flat roof areas surrounding the projecting monitors will give better and easier access to the glazed areas for maintenance and cleaning purposes.

The formation of the projecting monitor lights can be of light steelwork supported by lattice girders or standard universal beam sections; alternatively they can be constructed of cranked and welded universal beam sections supported on internal columns. To give long clear internal spans deep lattice beams of lightweight construction can be incorporated within the depth of the monitor framing in a similar manner to that used in northlight roofs — see Fig. VI.5. Precast concrete monitor portal frames can also be constructed for both single and multispan applications — see Fig. VI.4.

THERMAL INSULATION IN INDUSTRIAL BUILDINGS

The thermal insulation or resistance to the passage of heat in industrial buildings is covered by Building Regulation L1 and applies to industrial buildings having a floor area greater than 30 m^2 and which is likely to be heated by a space-heating system having an output exceeding 50 W/m^2 of floor area. The calculated heat loss for exposed floors, walls and roofs should not exceed 0.45 W/m^2 K with maximum windows areas of 15% of exposed wall area and rooflighting of 20% of roof areas. These percentages can be twice the figures quoted if double glazing is used and up to three times the permitted single glazing value if triple glazed or double glazed with a low emissivity coating.

The Approved Document supporting Building Regulation L1 gives alternative procedures to satisfy the requirements of the regulation and these are:

1. *Specified insulation thickness* — in this procedure the thermal conductivity of the insulating material (W/mK) must be known and by using the table in the Approved Document the base level thickness (mm) can be found. This base level thickness can be reduced by taking into account features of construction such as

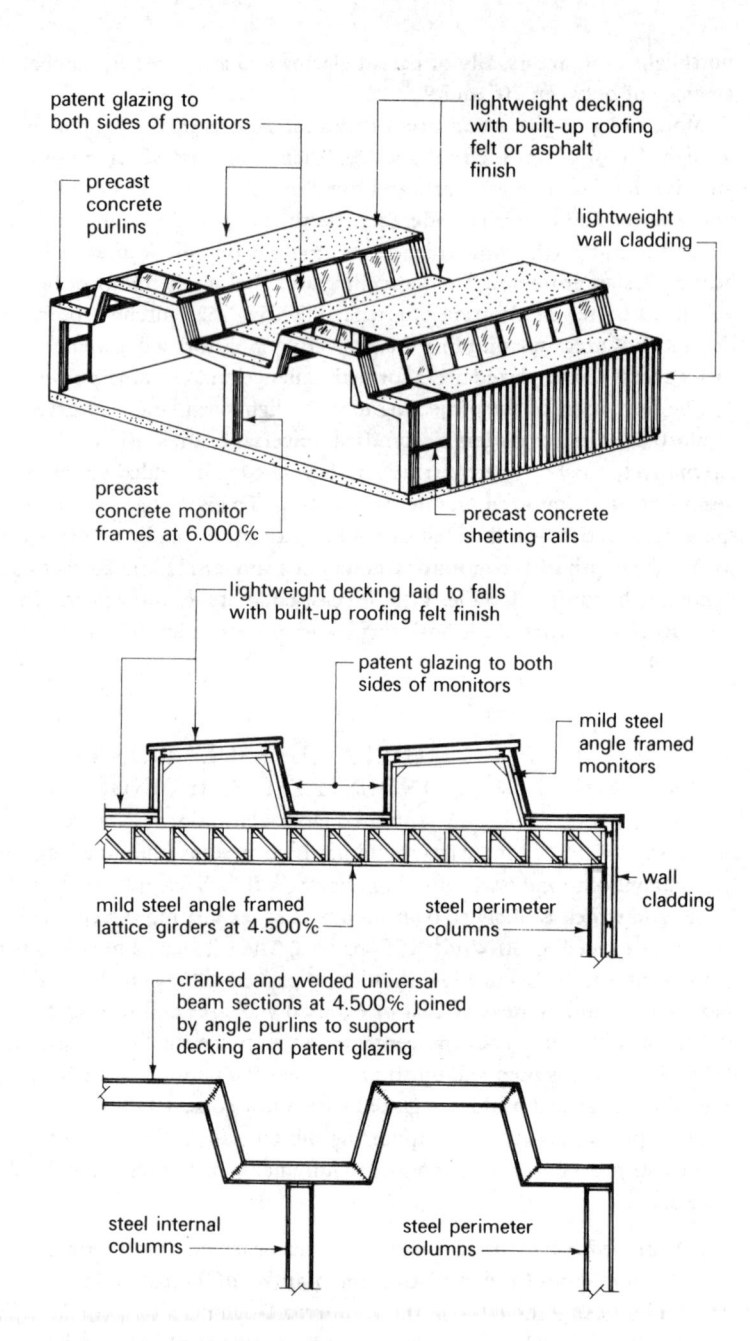

Fig VI.4 Typical monitor roof profiles

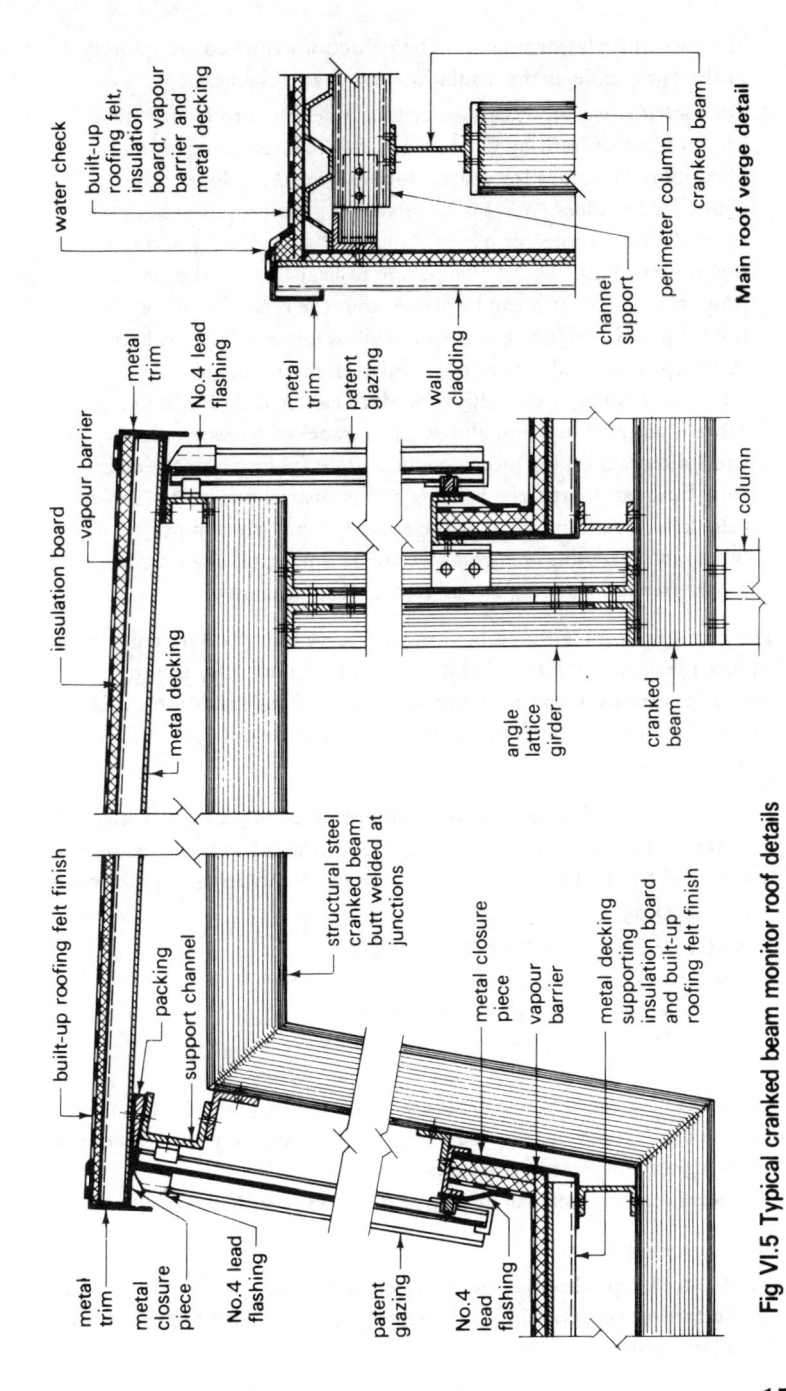

Fig VI.5 Typical cranked beam monitor roof details

air space and plaster finishes. The reductions allowed are quoted in the same table as the insulation base level thickness.
2. *Calculated trade off* – this procedure is designed to overcome the problem of limiting window and rooflight percentages and maximum 'U' values for walls, roofs and exposed floors. The allowable heat loss (W/K) is calculated for the proposed building using the limitations set out in Building Regulation L1 and this is compared with the calculated rate of heat loss for the actual proposed design. If it can be shown that the rate of heat loss in the proposed design is less than the allowable rate of heat loss the proposal would satisfy the regulation requirements.
3. *Calculated energy use* – this procedure can be used for buildings other than dwellings and allows a full trade-off between glazed and solid areas taking into account any useful heat gains such as solar heat gains, artificial lighting and industrial processes. The calculations must prove that the annual energy consumption taking into account heat gains would be no greater than if the values given in Building Regulation L1 had been used.

The heat loss from within a building is affected by the temperature difference between the internal and external environments, and to comply with various Acts and to create good working conditions, it is desirable to maintain a minimum temperature for various types of activities.

The ideal working temperature for any particular task is a subjective measure but as a guide the following internal temperatures recommended by the Institution of Heating and Ventilating Engineers are worth considering:

1. Sedentary work — 18.4°C minimum
2. Light work — 15.6°C minimum
3. Heavy work — 12.8°C minimum

The Factories Act 1961 requires a minimum temperature, after the first hour, of 15.6°C to be maintained except where sedentary work takes place where a higher temperature would be required.

The advantages which can be gained by having a well insulated roof are:

1. Lower fuel bills.
2. Reduced capital outlay on heating equipment.
3. Better working conditions for employees and hence better working relationships.

Sandwich construction

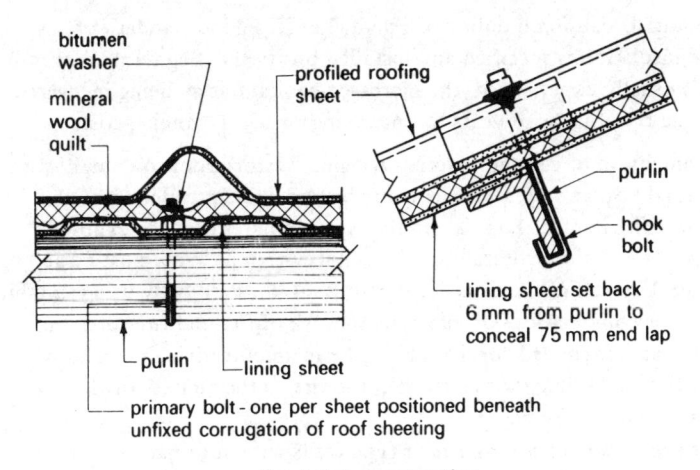

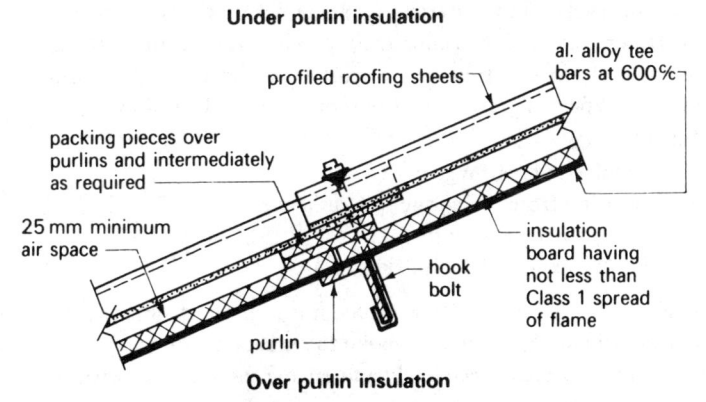

Under purlin insulation

Over purlin insulation

Fig VI.6 Factory roofs - typical insulation details

The initial cost of a building will be higher if a good standard of thermal insulation is specified and installed but in the long term an overall saving is usually experienced, the increased capital outlay being recovered within the first five or so years by the savings made in running costs.

The inclusion of certain materials within a factory roof to comply with the thermal insulation requirements of this Act may introduce into the structure a fire hazard. To this end the Act stipulates that the exposed surfaces of insulation materials used, even if within a cavity must be at least class 1 spread of flame as defined in BS 476 : Part 7. By using suitable materials or combinations of materials the risk of fire and fire spread in factories can be reduced considerably. The main objective, in common with all fires in buildings, is to contain the fire to the vicinity of the outbreak should a fire occur.

Factories can have compartment type walls with automatic closing fire-resistant doors as previously described (see Fig. VI.27) but if open planning is required other precautions will be necessary. The roof volume can be divided into cells by fitting permanent fire barriers within the triangulated profile of the roof structure with non-combustible materials such as fibre cement sheet suitably fire stopped within the profile of the roof covering. Beneath these fire barriers can be rolled curtains of fire fibre cloth controlled by a fusible link or similar fire detection device which will allow the curtain to fall forming a fire screen from roof to floor in the advent of a fire. A similar curtain could also be positioned in the longitudinal direction under a valley beam.

Using the above method a fire can be contained for a reasonable period within the confined area but it can also create another problem or hazard, that of smoke logging. The smoke generated by a fire will rise to the roof level and then start to circulate within the screened compartment, completely filling the volume of the confined section within a short space of time which apart from the hazard to people trying to escape can present the fire fighters with the following problems:

1. Difficulty in breathing.
2. Prevention from seeing the source of the fire.
3. Detecting nature of outbreak.
4. Assessing extent of outbreak.

A method of overcoming this problem is to have automatic high level ventilators which will allow the smoke to escape rapidly and thus give the fire fighters a chance to see clearly and enable them to deal with the outbreak. The use of ventilators, to overcome the problems of smoke logging, will of course introduce more air which aids combustion but

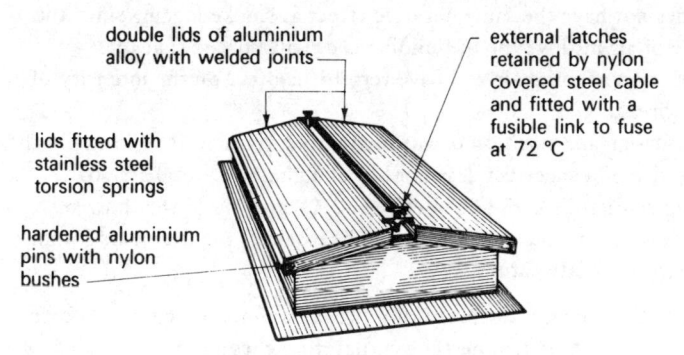

Double flap automatic fire ventilator

- double lids of aluminium alloy with welded joints
- lids fitted with stainless steel torsion springs
- hardened aluminium pins with nylon bushes
- external latches retained by nylon covered steel cable and fitted with a fusible link to fuse at 72 °C

- roof covering
- louvres in open position
- ridge capping
- base fixed under ridge capping at top edge
- filler pieces by main contractor
- ventilator of hardened aluminium alloy and nylon bushed pivots

Louvred fire ventilator

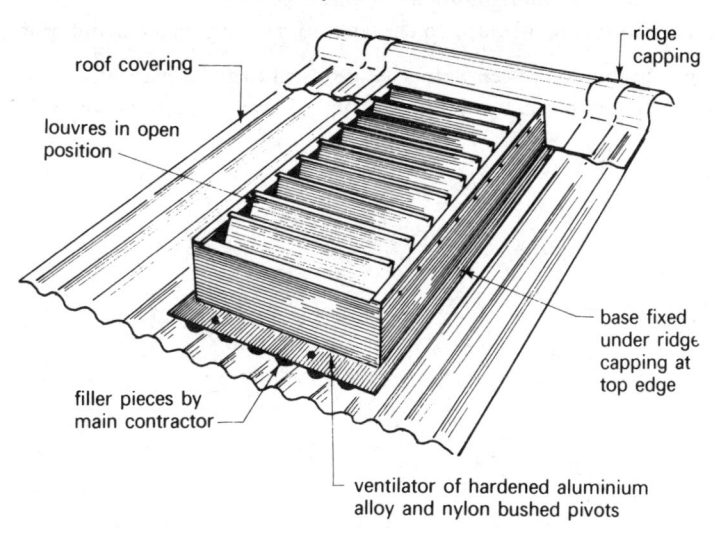

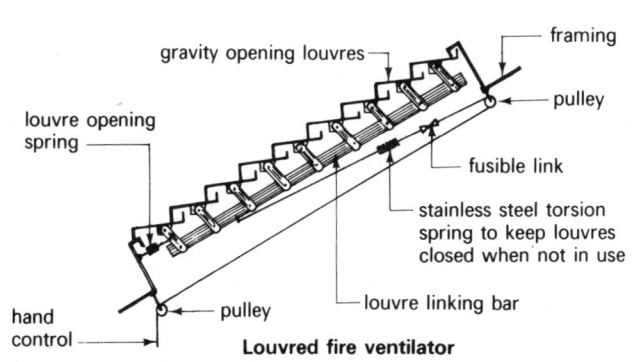

- gravity opening louvres
- framing
- pulley
- louvre opening spring
- fusible link
- stainless steel torsion spring to keep louvres closed when not in use
- hand control
- pulley
- louvre linking bar

Fig VI.7 Automatic roof fire ventilators

this does not have the same negative effect as smoke logging since the volume of air in this type of building is usually so vast that the introduction of more air will have very little effect on the intensity of the outbreak.

The design and position of automatic roof ventilators is normally the prerogative of a specialist designer but as a guide the total area of opening ventilators should be between 0.5% and 5% of the floor area depending on the likely area of fire. The essential requirements for an automatic ventilator are:

1. It must open in the event of a fire, a common specification being when the heat around the ventilator reaches a temperature of $68°C$.
2. Weatherproof under normal circumstances.
3. Easy to fix and blend with chosen roof covering material and profile.

Many automatic fire ventilators are designed to act as manually controlled ventilators under normal conditions — typical examples are shown in Fig. VI.7.

18
Walls

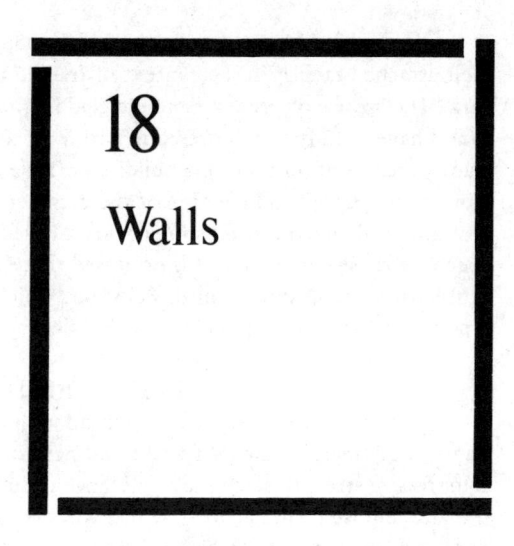

The walls of factory buildings have to fulfil the same functions as any enclosing wall to a building, namely:

1. Protection from the elements.
2. Provision of the required sound and thermal insulation.
3. Provision of the required degree of fire resistance.
4. Provide access to and exit from the interior.
5. Provide natural daylighting to the interior.
6. Give reasonable security protection to the premises.
7. Resist anticipated wind pressures.
8. Reasonable durability to keep long term maintenance costs down to an acceptable level.

Most contemporary factory buildings are constructed as framed structures using a three dimensional frame or a system of portal frames, which means that the enclosing walls can be considered as non-loading bearing claddings supporting their own dead weight plus any wind loading. The wall can be designed as a complete envelope masking entirely the structural framework using brick walling, precast concrete panels, curtain walling techniques or lightweight wall claddings or alternatively an infill panel technique could be used making a feature of the structural members.

The choice will depend on such factors as appearance, local planning requirements, short and long term costs and personal preference. Unless the factory is small, containing both works and offices within the same building and hence presenting the company's image to the would-be clients, appearance is very often considered to be of secondary importance.

Students should have already studied the topic of brick panel walls and their attached facings in the context of framed buildings (see Volume 2, Part III). The use of precast concrete cladding panels and lightweight infill panels have already been covered in Part V of this text. Many manufacturers of portal frame buildings offer a complete service of design, fabrication, supply and erection of the complete structure including the roof and wall coverings and students are advised to study the data sheets issued by these companies. It is proposed therefore to deal only with the lightweight wall claddings, in this chapter, which can be applied to the type of factory building under consideration.

LIGHTWEIGHT WALL CLADDINGS

In common with other cladding methods for framed buildings, lightweight wall claddings do not require high compressive strength since they only have to support their own dead load and any imposed wind loading, which will become more critical as the height and/or exposure increases. The subject of wind pressures is dealt with in greater detail in the next chapter. Lightweight claddings are usually manufactured from impervious materials which means that the run off of rain water can be high particularly under storm conditions when the discharge per minute could reach 2 litres per square metre of wall area exposed to the rain.

A wide variety of materials can be used as a cladding medium, most being profiled to a corrugated or trough form since the shaping will increase the strength of the material over its flat sheet form. Flat sheet materials are available but are rarely applied to large buildings because of the higher strength obtained from a profiled sheet of similar thickness. Special contoured sheets have been devised by many manufacturers to give the designer a wide range of choice in the context of aesthetic appeal. Claddings of various sandwich construction are also available to provide reasonable degrees of thermal insulation, sound insulation and to combat the condensation hazard which can occur with lightweight claddings of any nature.

The sheets are fixed in a similar manner to that studied in the second year of a typical construction technology course for sheet roof coverings. The support purlins are replaced in walls by a similar member called a sheeting rail which is fixed by cleats to the vertical structural frame members. The major difference occurs with the position of the fixings which in wall claddings are usually specified as being positioned in the trough of the profile as opposed to the crest when fixing roof coverings. This change in fixing detail is to ensure that the wall cladding is pulled tightly up to the sheeting rail or lining tray.

Plastic protective caps for the heads of fixings are available, generally of a colour and texture which will blend with the wall cladding. A full range of fittings and trims are usually obtainable for most materials and profiles to accommodate openings, returns, top edge and bottom edge closing. Typical cladding details are shown in Figs. VI.8 and VI.9.

Common materials used for lightweight wall claddings are:

1. *Fibre cement* — non-combustible material in corrugated and troughed sheets which are generally satisfactory when exposed to the weather but are susceptible to impact damage. Average life is about 20 years which can be increased considerably by paint protection. Unpainted sheets loose their surface finish at the exposed surface by carbonation and become ingrained with dirt. To achieve reasonable thermal insulation standards a lining material will be required, which is normally sandwiched between the cladding and an inner lining tray.

2. *Coated steel sheets* — non-combustible material with a wide range of profiles produced by various manufacturers. The steel sheet forms the core of the cladding providing its strength and this is covered with various forms of coatings to give weather protection, texture and colour. A typical specification would be a galvanised steel sheet core covered on both sides with a layer of asbestos felt to increase resistance to fire, a layer of bitumen-impregnated felt to act as a barrier to the passage of moisture to the core and on the face surface a coloured and textured coating of plastic. Fixing and the availability of fittings is as described above for asbestos cement.

3. *Aluminium alloy sheets* — non-combustible material in corrugated and troughed profiles which are usually made to the recommendations of CP 143 Pt 1 and BS 4868 respectively. Other profiles are also available as manufacturers' standards. Durability will depend upon the alloy used but this can be increased by paint applications; if unpainted, regular cleaning may be necessary if its natural bright appearance is to be maintained. Fixing, fittings and the availability of linings is as given for other cladding materials.

4. *Polyvinyl chloride sheets* — generally supplied in a corrugated profile with an embedded wire reinforcement to provide a cladding with a surface spread of flame classification of class 1 in accordance with BS 476: Part 7. The durability of this form of cladding is somewhat lower than those previously considered and the colours available are limited. The usual range of fittings and trims are available.

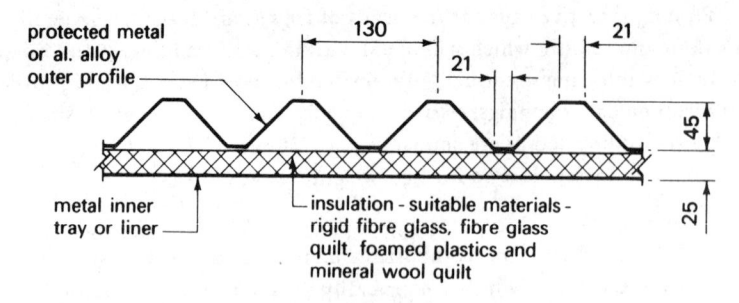

Typical cladding profile

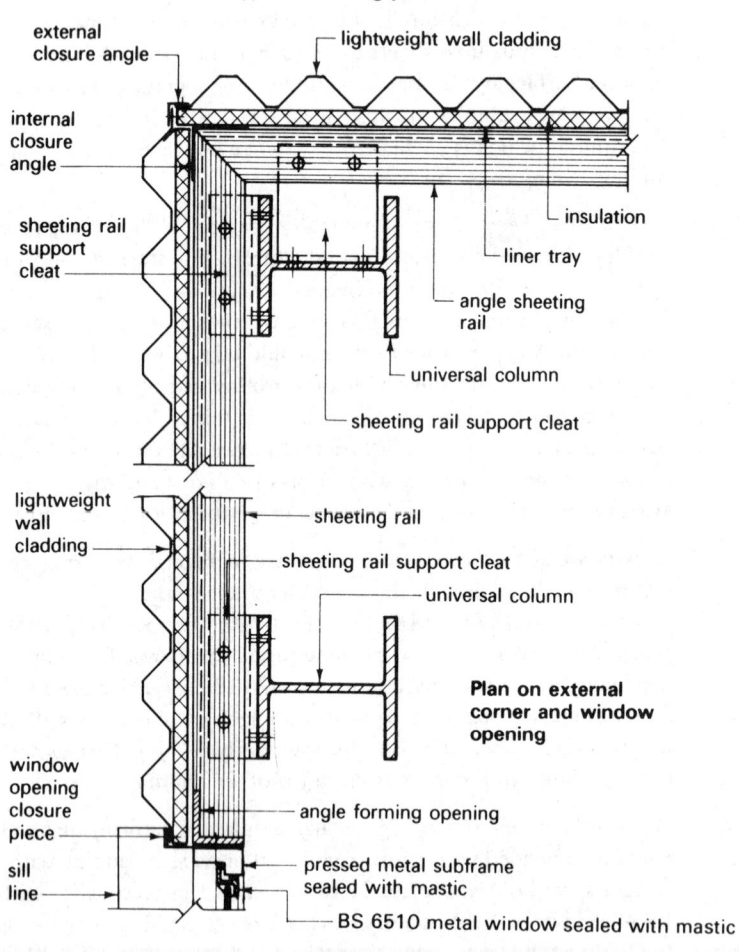

Fig VI.8 Lightweight wall cladding - typical details 1

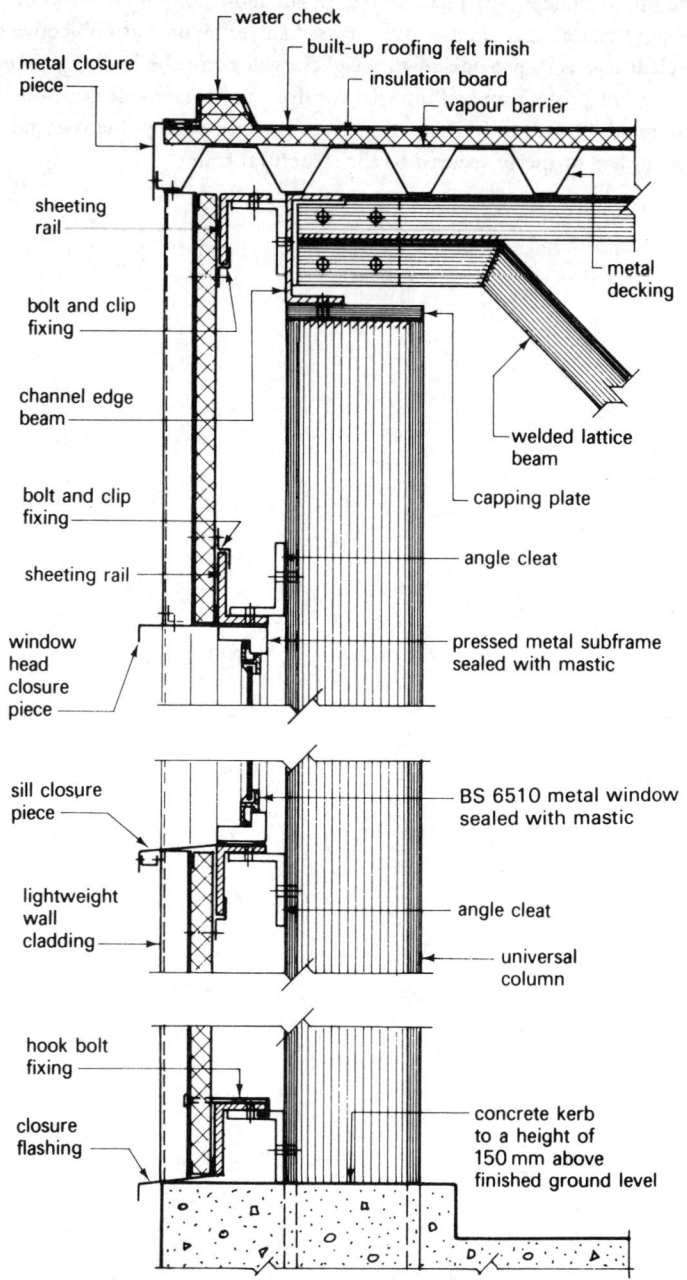

Fig VI.9 Lightweight wall cladding - typical details 2

The importance of adequate design, detail and fixing of all forms of lightweight cladding cannot be overstressed since the primary objective of these claddings is to provide a lightweight envelope to the building giving basic weather protection and internal comfort at a reasonable cost. Claddings which will fulfil these objectives are very susceptible to wind damage unless properly secured to the structural frame.

19
Wind pressures

Wind can be defined as a movement of air, the full nature of which is not fully understood, but two major contributory factors which can be given are:

1. Convection currents caused by air being warmed at the earth's surface, becoming less dense, rising and being replaced by colder air.
2. Transference of air between high and low pressure areas.

The speed with which the air moves in replacement or transfer is termed its velocity and can be from 0 to 1.5 metres per second, when it is hardly noticeable, to speeds in excess of 24 metres per second, when considerable damage to property and discomfort to persons could be the result.

The physical nature of the ground or topography over which the wind passes will have an effect on local wind speeds since obstructions such as trees and buildings can set up local disturbances by forcing the wind to move around the sides of the obstruction or funnel between adjacent obstacles. Where funnelling occurs the velocity and therefore the pressure can be increased considerably. Experience and research has shown that the major damage to buildings is caused not by a wind blowing at a constant velocity but by the short duration bursts or gusts of wind of greater intensity than the prevailing mean wind speed. The durations of these gusts are usually measured in 3, 5 and 15 second periods and information of the likely maximum gust speeds for specific long term time durations of 50 years or more are available from the Meteorological Office.

When the wind encounters an object in its path such as the face of a building it is usually rebuffed and forced to turn back on itself, this has the effect of setting up a whirling motion or eddy which eventually finds its way around or over the obstruction. The pressure of the wind is normally in the same direction as the path of the wind which tends to push the wall of the building inwards, and indeed will do so if sufficient resistance is not built into the structure. The effect of local eddies however is very often opposite in direction and force to that of the prevailing wind, producing a negative or suction force — see Fig. VI.10.

Many factors must be taken into account before the magnitude and direction of wind pressures can be determined; these include height to width ratio of the building, length to width ratio of the building, plan shape of the building, approach topography, exposure of the building and the proximity of surrounding structures. Account must also be taken of any likely openings in the building, since the entry of wind will exert a positive pressure on any walls or ceilings encountered. These internal pressures must be added to or subtracted from the type of pressure anticipated acting on the external face at the same point. For a detailed study of the effects and assessment of wind loads on buildings students are advised to examine the contents of BRE Digests numbers 119 and 141.

All buildings are at some time subjected to wind pressures but some are more vulnerable than others due to their shape, exposure or method of construction. One method of providing adequate resistance to wind pressures is to use materials of high density; it follows therefore that buildings which are clad with lightweight coverings are more susceptible to wind damage than those using the heavier traditional materials. Factory buildings using lightweight claddings have therefore been taken to serve as an illustration of providing suitable means of resistance to wind pressures.

To overcome the problem of uplift or suction on roofs caused by the negative wind pressures adequate fixing or anchorage of the lightweight coverings to the structural frame is recommended. Generally sufficient resistance to uplift of the frame is inherent in the material used for the structural members, the problem is therefore to stop the covering being pulled away from the supporting member. This can be achieved by the quality or holding power of the fixings used or by the number of fixings employed or by a combination of both. It should be noted that the whole roof considered as a single entity is at risk and not merely individual sheets.

If the supporting member does not have sufficient self dead load to overcome the suction forces, such as a timber plate bedded on to a brick wall, then it will be necessary adequately to anchor the plate to the wall. This can be carried out by means of bolts or straps fixed to the plate and

Low buildings

pitched roof of less than 30°
suction
severe suction
eddy
pressure

Notes: roofs over 35° pitch usually develop positive pressure on windward slope with suction over ridge and on lee slope — wind in direction of ridge will create areas of suction along all windward edges

pressure on windward face

suction and acceleration around ends

large eddy exerting suction on rear face

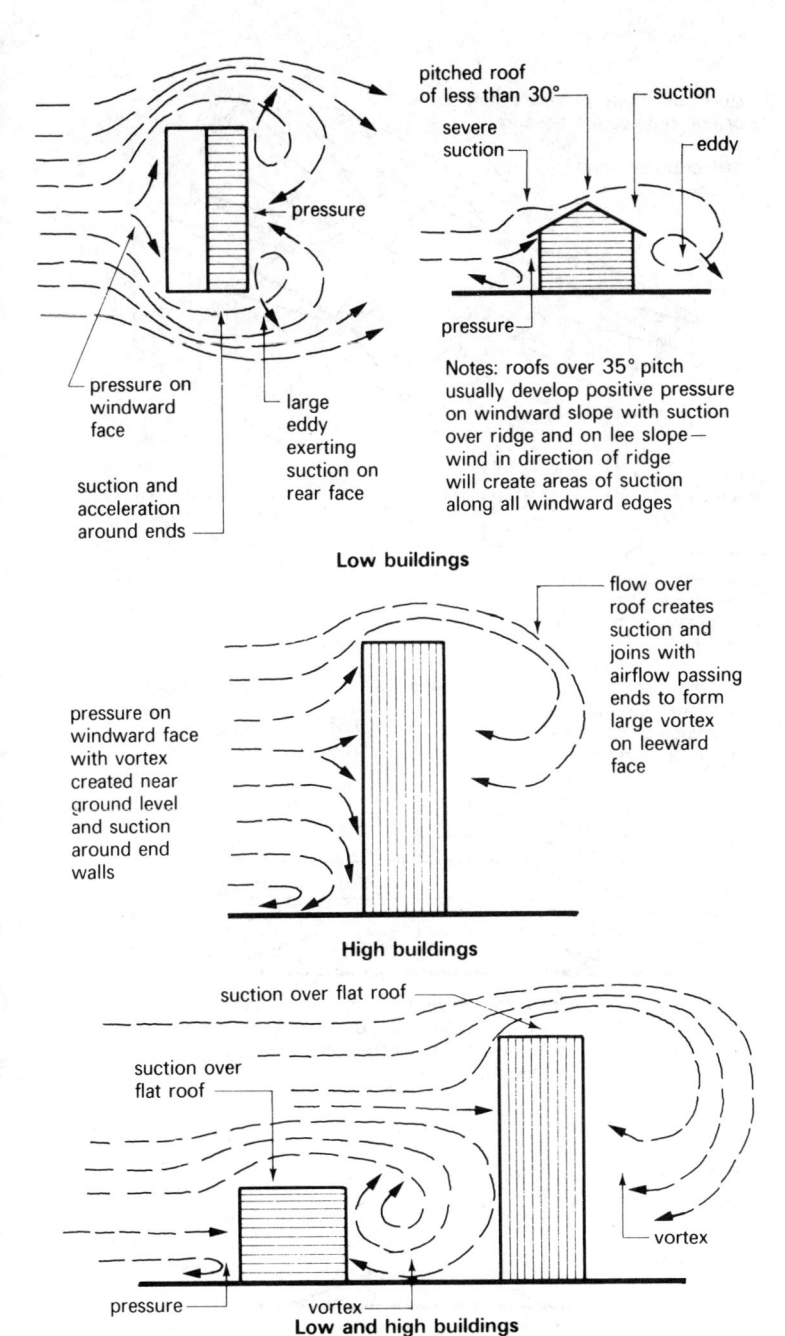

pressure

High buildings

pressure on windward face with vortex created near ground level and suction around end walls

flow over roof creates suction and joins with airflow passing ends to form large vortex on leeward face

Low and high buildings

suction over flat roof

suction over flat roof

pressure

vortex

vortex

Fig VI.10 Typical wind pressures around buildings

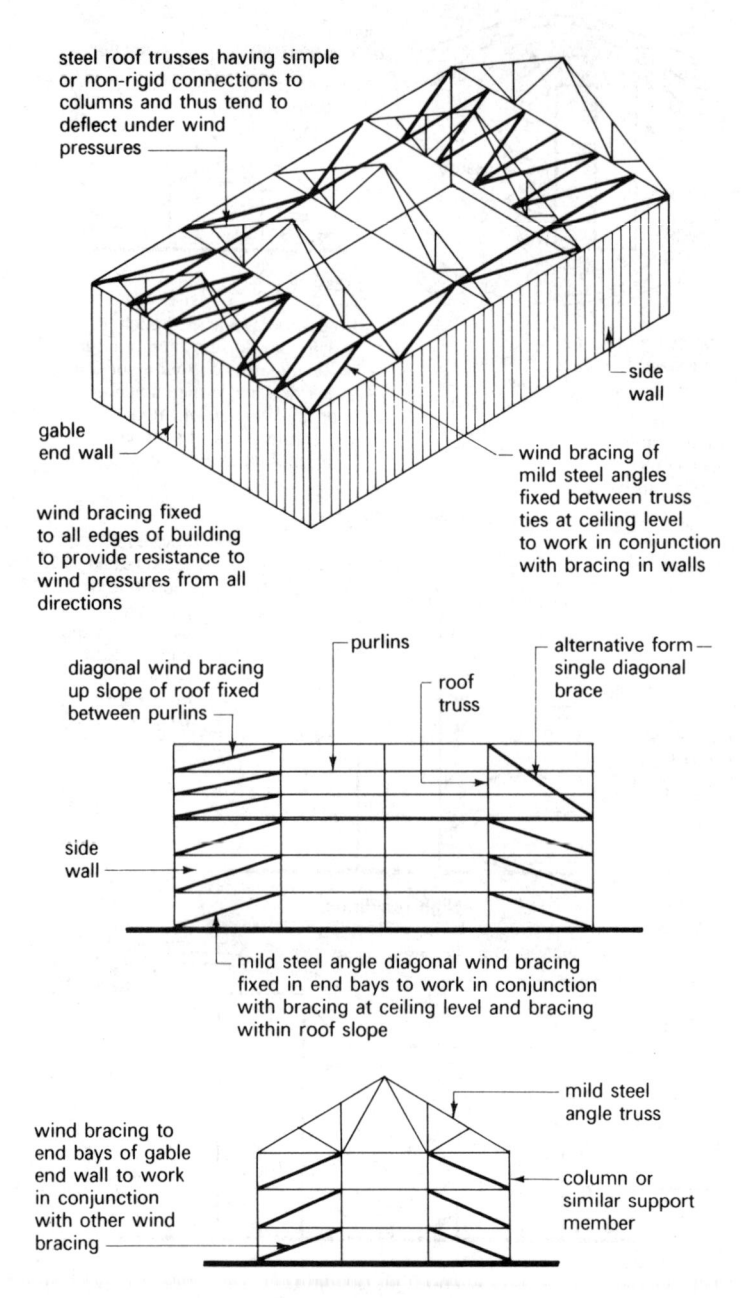

Fig VI.11 Typical wind bracing arrangements

wall in such a manner that part of the dead load of the wall can be added to that of the supporting member.

Positive wind pressures tend to move or bend the wall forwards in the same direction of the wind; this tendency is usually overcome when using light structural framing by adding stiffeners called wind bracing to the structure. Wind braces are usually of steel angle construction fitted between the structural members where unacceptable pressures are anticipated. They take the form of cross bracing, forming what is in fact a stiffening lattice within the frame — see Fig. VI.11. Although each area of the country has a predominant prevailing wind buildings requiring wind bracing are usually treated the same at all likely vulnerable positions to counteract changes in wind direction and the effect of local eddies.

The immense destructive power of the wind, in the context of building works, cannot be over emphasised and careful consideration is required from design stage to actual construction on site. Students should also appreciate that temporary works and site hutments are just as vulnerable to wind damage as the finished structure. Great care must be taken therefore when planning site layouts, plant positioning, erection of scaffolds and hoardings if safe working conditions are to be obtained on building sites.

Part VII
Formwork

20
Wall formwork

Formwork is a temporary mould into which wet concrete and reinforcement is placed to form a particular desired shape with a predetermined strength. Depending upon the complexity of the form, the relative cost of formwork to concrete can be as high as 75% of the total cost to produce the required member. A typical breakdown of percentage costs could be as follows:

Concrete	materials	28%	40%
	labour	12%	
Reinforcement	materials	18%	25%
	labour	7%	
Formwork	materials	15%	35%
	labour	20%	

The above breakdown shows that a building contractor will have to use an economic method of providing the necessary formwork if he is to be competitive in tendering since this is the factor over which he has most control.

The economic essentials of formwork can be listed thus:—

1. *Low cost* — only that amount of money necessary, which will produce the required form, to be expended.
2. *Strength* — careful selection of formwork materials and support members to obtain the most economic balance in terms of quantity used and continuing site activity around the assembled formwork.

3. *Finish* — selection of method, materials and if necessary linings to produce the desired result direct from the formwork. Applied finishes are usually specified and therefore only method is the real factor over which the builder would have any economic control.
4. *Assembly* — consideration must be given to the use of patent systems and mechanical handling plant.
5. *Material* — advantages of using either timber or steel should be considered; generally timber is lighter in weight and therefore larger units could be used, but steel will give more uses than timber although it cannot be repaired as easily.
6. *Design* — within the confines of the architectural and/or structural design formwork should be as repetitive and adaptable as possible.

A balance of the above essentials should be achieved, preferably at pretender stage, so that an economic and competitive cost can be calculated.

By the time the student has reached the first year of advanced construction technology he should have already studied the basic principles of formwork and in particular formwork for columns, beams and slabs. It is a useful exercise at this stage to recapitulate on these fundamentals (see Volume 2, Part III) before proceeding with a study of wall formwork and patent systems.

In principle the design, fabrication and erection of wall formwork is similar to that already studied in the context of column formwork. Several basic methods are available which will enable a wall to be cast in large quantities, defined lifts or continuous from start to finish.

TRADITIONAL WALL FORMWORK

This usually consists of standard framed panels tied together over their backs with horizontal members called walings. The walings fulfil the same function as the yokes or column clamps of providing the resistance to the horizontal force of wet concrete. A 75 mm high concrete kicker is formed at the base of the proposed wall to enable the forms to be accurately positioned and to help prevent the loss of grout by seepage at the base of the form.

The usual assembly is to erect one side of the wall formwork ensuring that it is correctly aligned, plumbed and strutted. The steel reinforcement cage is inserted and positioned before the other side formwork is erected and fixed. Keeping the forms parallel and at the correct distance from one another is most important; this can be achieved by using precast concrete spacer blocks which are cast in, steel spacer tubes which are removed after casting and curing, the voids created being made good, or alternatively by using one of the many proprietary wall tie spacers available — see Fig. VII.1

To keep the number of ties required within acceptable limits horizontal members or walings are used, these also add to the overall rigidity of the formwork panels and help with alignment. Walings are best if composed of two members with a space between which will accommodate the shank of the wall tie bolt; this will give complete flexibility in positioning the ties and leave the waling timbers undamaged for eventual re-use. To ensure that the loads are evenly distributed over the pair of walings plate washers should be specified.

Plywood sheet is the common material used for wall formwork but this material is vulnerable to edge and corner damage. The usual format is therefore to make up wall forms as framed panels on a timber studwork principle with a plywood facing sheet screwed to the studs so that it can be easily removed and reversed to obtain the maximum number of uses.

Corners and attached piers need special consideration since the increased pressures at these points could cause the abutments between panels to open up, giving rise to unacceptable grout escape and a poor finish to the cast wall. The walings can be strengthened by including a loose tongue at the abutment position and extra bracing could be added to internal corners — see Fig. VII.2. When considering formwork for attached piers it is usually necessary to have special panels to form the reveals.

CLIMBING FORMWORK

This is a method of casting a wall in set vertical lift heights using the same forms in a repetitive fashion thus obtaining maximum usage from a minimum number of forms. The basic formwork is as shown in Fig. VII.3, which in the first lift is positioned against the kicker in the inverted position, the concrete is poured and allowed to cure after which the forms are removed, reversed and fixed to the newly cast concrete. After each casting and curing of concrete the forms are removed and raised to form the next lift until the required height has been reached.

It is possible to use this method for casting walls against an excavated or sheet piled back face using formwork to one side only by replacing the through wall tie spacers with loop wall ties — see Fig. VII.3.

When using this single sided method adequate bracing will be required to maintain the correct wall thickness and when the formwork is reversed, after the first lift, it must be appreciated that the uprights or soldiers are acting as cantilevers and will therefore need to be stronger than those used in the double-sided version.

Component parts of coil ties

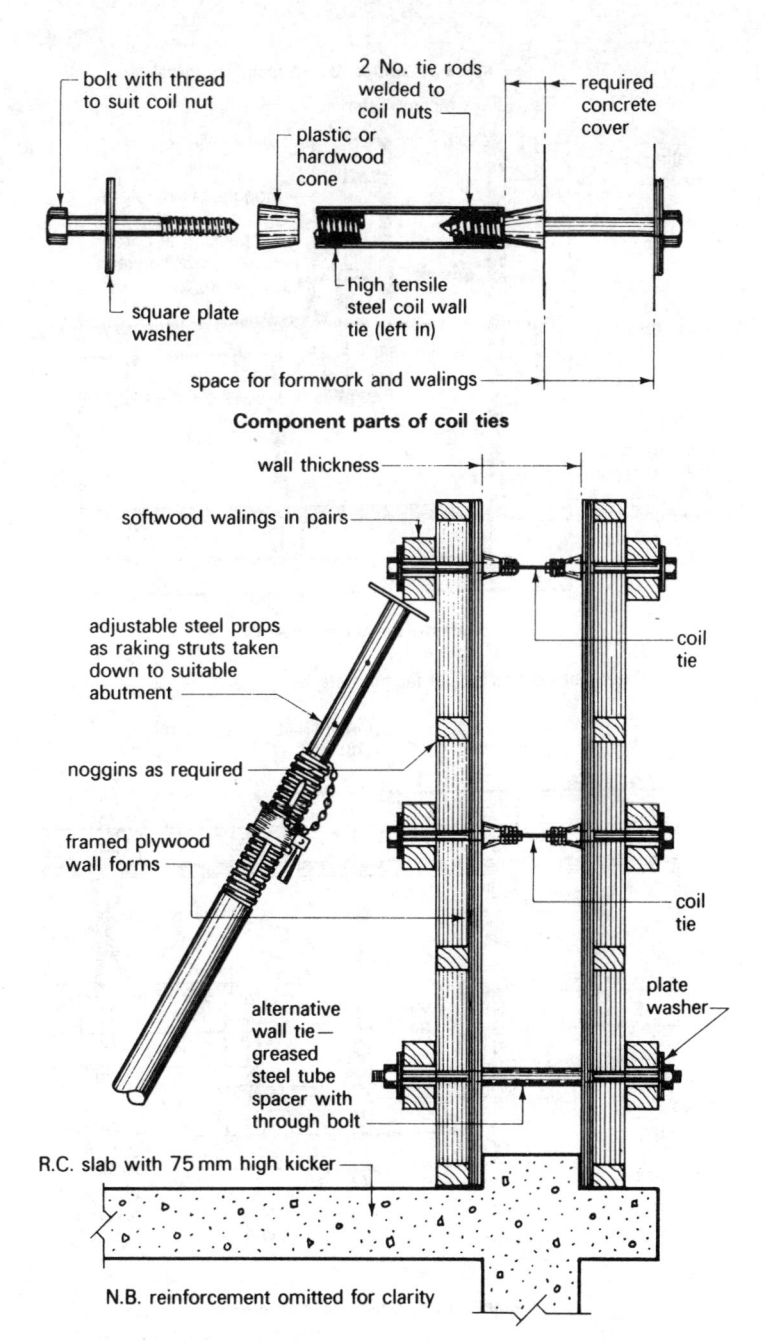

Fig VII.1 Traditional wall formwork - details 1

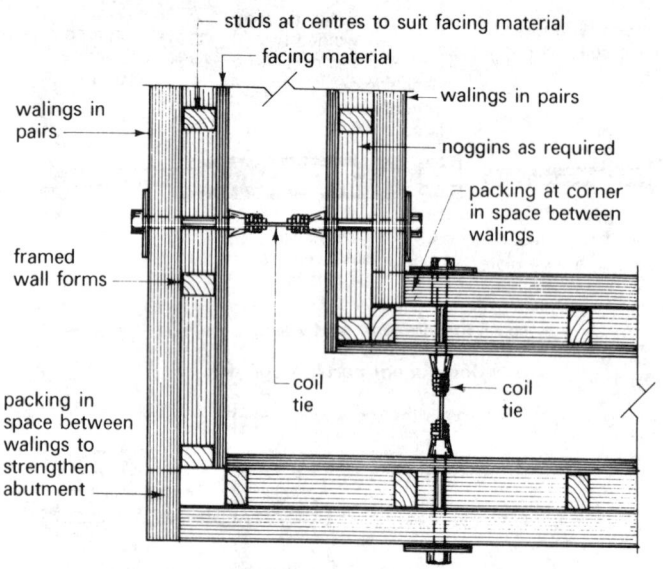

Plan on corner formwork

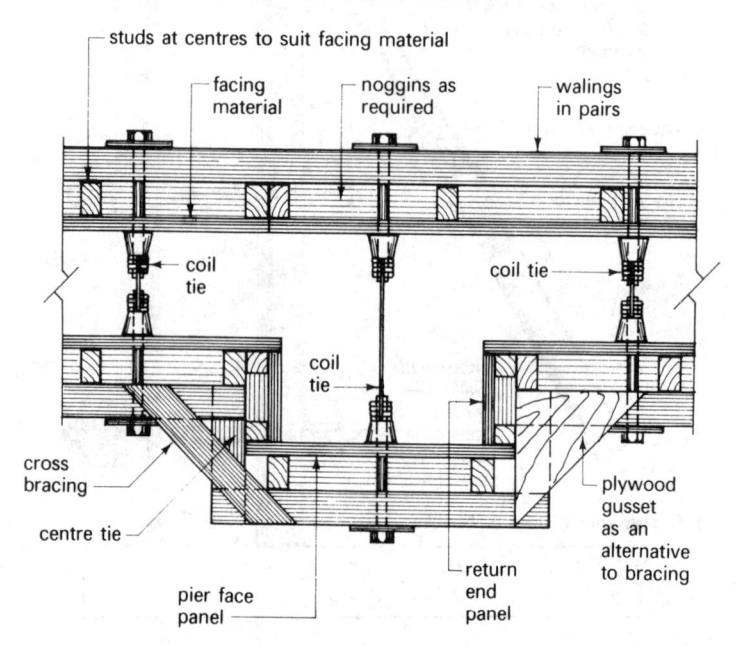

Plan on attached pier formwork

Fig VII.2 Traditional wall formwork - details 2

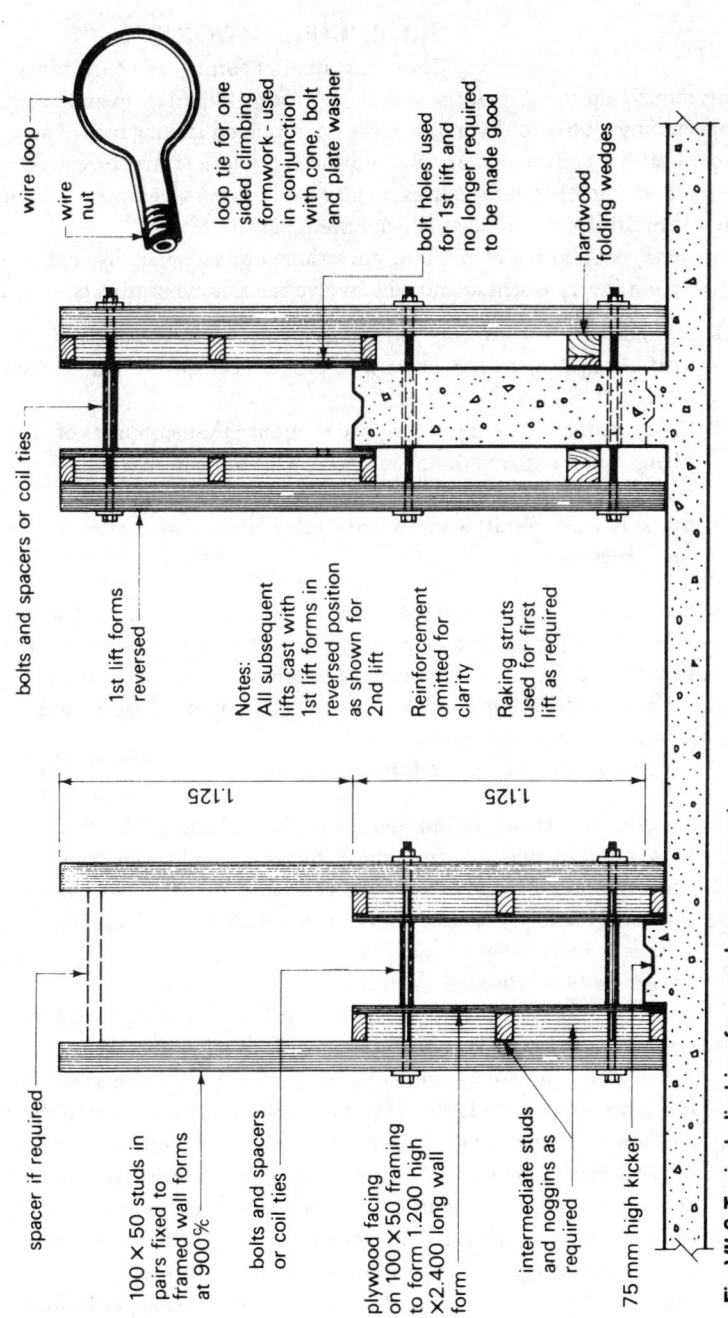

Fig VII.3 Typical climbing formwork arrangement

SLIDING FORMWORK

This is a system of formwork which slides continuously up the face of the wall being cast by climbing up and being supported by a series of hydraulic jacks operating on jacking rods. The whole wall is therefore cast as a monolithic and jointless structure making the method suitable for structures such as water towers, chimneys and the cores of multi-storey buildings which have repetitive floors.

Since the system is a continuous operation good site planning and organisation is very essential and will involve the following aspects:

1. Round-the-clock working which will involve shift working and artificial lighting to enable work to proceed outside normal daylight hours.
2. Careful control of concrete supply to ensure that stoppages of the lifting operation are not encountered. This may mean having standby plant as an insurance against mechanical breakdowns.
3. Suitably trained staff accustomed to this method of constructing *in situ* concrete walls

The actual architectural and structural design must be suitable for the application of a slipform system; generally the main requirements are a wall of uniform thickness with a minimum number of openings and a height of at least 20.000 m to make the cost of equipment, labour and planning an economic proposition.

The basic components of slip formwork are:

1. *Side forms* — these need to be strongly braced and are loadbearing of timber and/or steel construction. Steel forms are heavier than timber, more difficult to assemble and repair but they have lower frictional loading, are easier to clean and have better durability. Timber forms are lighter, have better flexibility, easier to repair and are generally favoured. A typical timber form would consist of a series of 100 × 25 planed straight grained staves assembled with a 2 mm wide gap between consecutive boards to allow for swelling which could give rise to unacceptable friction as the forms rise. The forms are usually made to a height of 1.200 m with an overall sliding clearance of 6 mm by keeping the external panel plumb and the internal panel tapered so that it is 3 mm in at the top and 3 mm out at the bottom, giving the true wall thickness, in the centre position of the form. The side forms must be adequately stiffened with horizontal walings and vertical puncheons to resist the lateral pressure of concrete and transfer the loads of working platforms to the supporting yokes.

2. *Yokes* — assist in supporting the suspended working platforms and transfers the platform and side form loads to the jacking rods. Yokes are usually made of framed steelwork suitably braced and designed to provide the necessary bearings for the working platforms.
3. *Working platforms* — three working levels are usually provided, the first is situated above the yokes at a height of about 2.000 m above the top of the wall forms for the use of the steel fixers. The second level is a platform over the entire inner floor area at a level coinciding with the top of the wall forms and is used by the concrete gang, for storage of materials, to carry levelling instruments and jacking control equipment. It is worth noting that this decking could ultimately be used as the soffit formwork to the roof slab if required. The third platform is in the form of a hanging or suspended scaffold usually to both sides of the wall and is to give access to the exposed freshly cast concrete below the slip formwork for the purpose of finishing operations.
4. *Hydraulic jacks* — the jacks used are usually specified by their load bearing capacities such as 3 tonnes and 6 tonnes and consist of two clamps operated by a piston. The clamps operate on a jacking rod of 25 to 50 mm diameter according to the design load and are installed in banks operated from a central control to give an all round consistent rate of climb. The upper clamp grips the jacking rod and the lower clamp, being free, rises, pulling the yoke and platforms with it until the jack extension has been closed. The lower clamp now grips the climbing rod whilst the upper clamp is released and raised to a higher position when the lifting cycle is recommenced. Factors such as temperature and concrete quality affect the rate of climb but typical speeds are between 150 and 450 mm per hour.

The upper end of the jacking rod is usually encased in a tube or sleeve to overcome the problem of adhesion between the rod and the concrete. The jacking rod therefore remains loose in the cast wall and can be recovered at the end of the jacking operation. The 2.500 to 4.000 m lengths of rod are usually joined together with a screw joint arranged so that all such joints do not occur at the same level. A typical diagrammatic arrangement of sliding formwork is shown in Fig. VII.4.

The site operations commence with the formation of a substantial kicker 300 mm high incorporating the wall and jacking rod starter bars. The wall forms are assembled and fixed together with the yokes, upper working platforms and jacking arrangement after which the initial concrete lift is poured. The commencing rate of climb must be slow to allow time for the first batch of concrete to reach a suitable condition before emerging from beneath the sliding formwork. A standard or planned rate

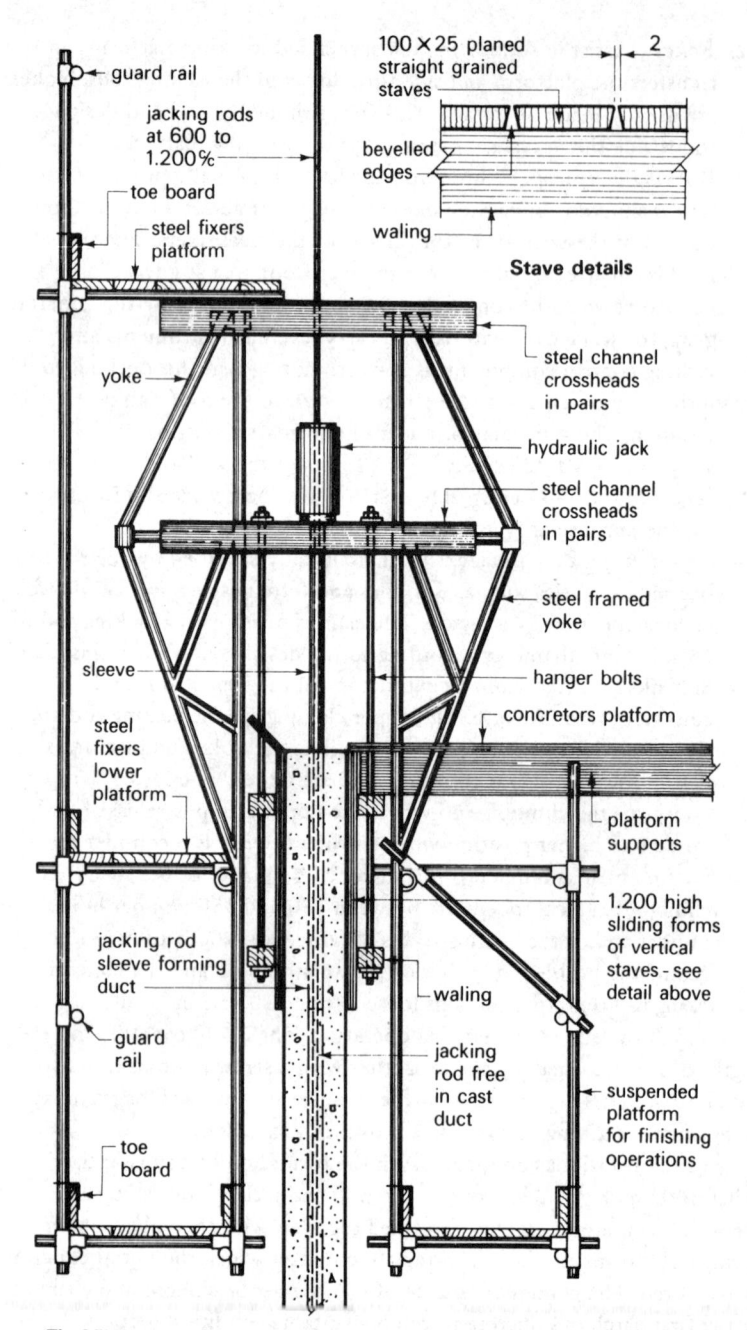

Fig VII.4 Diagrammatic arrangement of sliding formwork

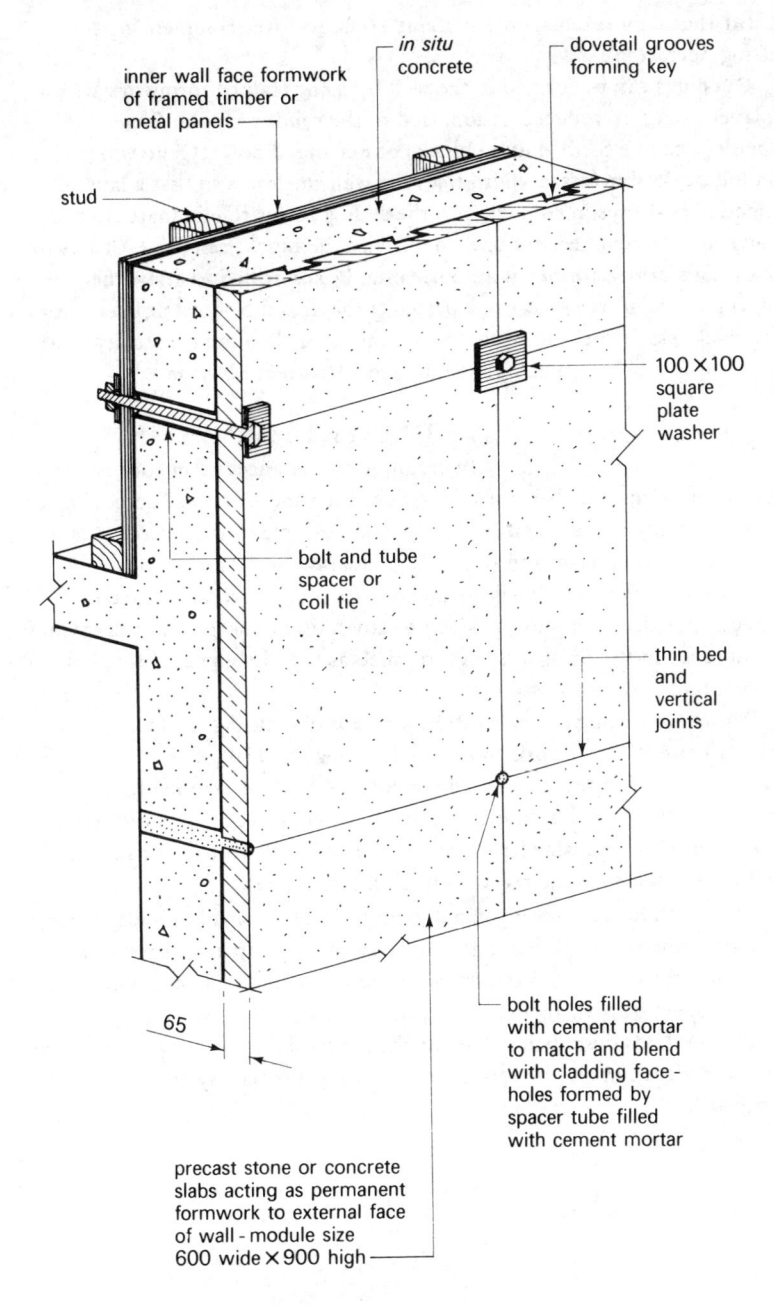

Fig VII.5 Example of permanent formwork

of lift is usually reached within about 16 hours after commencing the lifting operation.

Openings can be formed in the wall by using framed formwork with splayed edges, to reduce friction, tied to the reinforcement. Small openings can be formed using blocks of expanded polystyrene which should be 75 mm less in width than the wall thickness so that a layer of concrete is always in contact with the sliding forms to eliminate friction. The concrete cover is later broken out and the blocks removed. Chases for floor slabs can be formed with horizontal boxes drilled to allow the continuity reinforcement to be passed through and to be bent back within the thickness of the wall so that when the floor slabs are eventually cast the reinforcement can be pulled out into its correct position.

PERMANENT FORMWORK

In certain circumstances formwork is left permanently in place because of the difficulty and/or cost of removing it once the concrete has been cast. A typical example of these circumstances would be when a beam and slab raft foundation with shallow upstand beams and an *in situ* slab have been constructed. Apart from the cost aspect, consideration must be given to any nuisance such an arrangement could cause in the finished structure, such as the likelihood of fungi or insect attack and the possible risk of fire.

Permanent formwork can also be a means of utilising the facing material as both formwork and outer cladding especially in the construction of *in situ* reinforced concrete walls. The external face or cladding is supported by the conventional internal face formwork, which can in certain circumstances overcome the external strutting or support problems often encountered with high rise structures.

This method is however generally limited to thin small modular facing materials, the size of which is governed by the supporting capacity of the internal formwork. Fig. VII.5 shows a typical example of the application of this aspect of permanent formwork.

The methods described above for the construction of *in situ* reinforced concrete walls can also be carried out by using a patent system of formwork.

21
Patent formwork

Patent formwork is sometimes called system formwork and is usually identified by the manufacturer's name. All proprietary systems have the same common aim and most are similar in their general approach to solving some of the problems encountered with formwork for modular designed or repetitive structures. As shown in the previous chapter, formwork is one area where the contractor has most control over the method and materials to be used in forming an *in situ* reinforced concrete structure. In trying to design or formulate the ideal system for formwork the following must be considered:

1. *Strength* — to carry the concrete and working loads.
2. *Lightness without strength reduction* — to enable maximum size units to be employed.
3. *Durability without prohibitive costs* — to gain maximum usage of materials.
4. *Good and accurate finish straight from the formwork* — to reduce the costly labour element of making good and patching, which in itself is a difficult operation to accomplish without it being obvious that this kind of treatment was found necessary.
5. *Erection and dismantling times.*
6. *Ability to employ unskilled or semi-skilled labour to carry out the work.*

Patent or formwork systems have been devised to satisfy most of the above listed requirements by the standardisation of forms and by easy methods of securing and bracing the positioned formwork.

The major component of any formwork system is the unit panel which should fulfil the following requirements:

1. Available in a wide variety of sizes based on a standard module, usually multiples and submultiples of 300 mm.
2. Manufactured from durable materials.
3. Covered with a facing material which is durable and capable of producing the desired finish.
4. Units should be interchangeable so that they can be used for beams, columns and slabs.
5. Formed so that they can be easily connected together to form large unit panels.
6. Lightweight so that individual unit panels can be handled without mechanical aid.
7. Designed so that the whole formwork can be assembled and dismantled easily by unskilled or semi-skilled labour.
8. Capable of being adapted so that non-standard width inserts of traditional formwork materials can be included where lengths or widths are not exact multiples of the unit panels.

Most unit panels consist of a framed tray made from light metal angle or channel sections stiffened across the width as necessary. The edge framing is usually perforated with slots or holes to take the fixing connectors and waling clips or clamps. The facing is of sheet metal or plywood, some manufacturers offering a choice. Longitudinal stiffening and support is given by clamping, over the backs of the assembled panels, special walings of hollow section or in many systems standard scaffold tubes are used. Vertical support where required can be given by raking standard adjustable steel props, or where heavy loadings are encountered most systems have special vertical stiffening and support arrangements based on designed girder principles which also provide support for an access and working platform. Spacing between opposite forms is maintained using wall ties or similar devices as previously described for traditional formwork.

Walls which are curved in plan can be formed in a similar manner to the straight wall techniques described above using a modified tray which has no transverse stiffening members thus making it flexible. Climbing formwork can also be carried out using system formwork components but instead of reversing the forms as described for traditional formwork climbing shoes are bolted to the cast section of the wall to act as bearing corbels to supports the soldiers for each lift.

When forming beams and column the unit panels are used as side or soffit forms held together with steel column clamps, or in the case of beams conventional strutting or alternatively using wall ties through the beam thickness. Some manufacturers produce special beam box clamping devices which consist of a cross member surmounted by attached and adjustable triangulated struts to support the side forms.

Many patent systems for the construction of floor slabs which require propping during casting use basic components of unit panels, narrow width (150 mm) filler panels, special drop head adjustable steel props, joists and standard scaffold tubes for bracing. The steel joists are lightweight, purpose made and are supported on the secondary head of the prop in the opposite direction to the filler panels which are also supported by the secondary or drop head of the prop. The unit panels are used to infill between the filler panels and pass over the support joist, the upper head of the prop being at the same level as the slab formwork and is indeed part of the slab soffit formwork.

After casting the slab and allowing it to gain sufficient strength the whole of the slab formwork can be lowered and removed leaving the undisturbed prop head to give the partially cured slab a degree of support. This method enables the formwork to be removed at a very early stage releasing the unit panels, filler pieces and joists for re-use and at the same time accelerating the drying out of the concrete by allowing a free air flow on both sides of the slab.

Slabs of moderate spans (up to 7.500 m) which are to be cast between load bearing walls or beams without the use of internal props can be formed using unit panels supported by steel telescopic floor centres. These centres are made in a simple lattice form extending in one or two directions according to span and are light enough to be handled by one man. They are precambered to compensate any deflections when loaded. Typical examples of system formwork are shown in Figs. VII.6 to VII.9.

TABLE FORMWORK

This special class of formwork has been devised for use when casting large repetitive floor slabs in medium to high rise structures. The main objective is to reduce the time factor in erecting, striking and re-erecting slab formwork by creating a system of formwork which can be struck as an entire unit, removed, hoisted and repositioned without any dismantling.

The basic requirements for a system of table formwork can be listed thus:

1. A means of adjustment for aligning and levelling the forms.

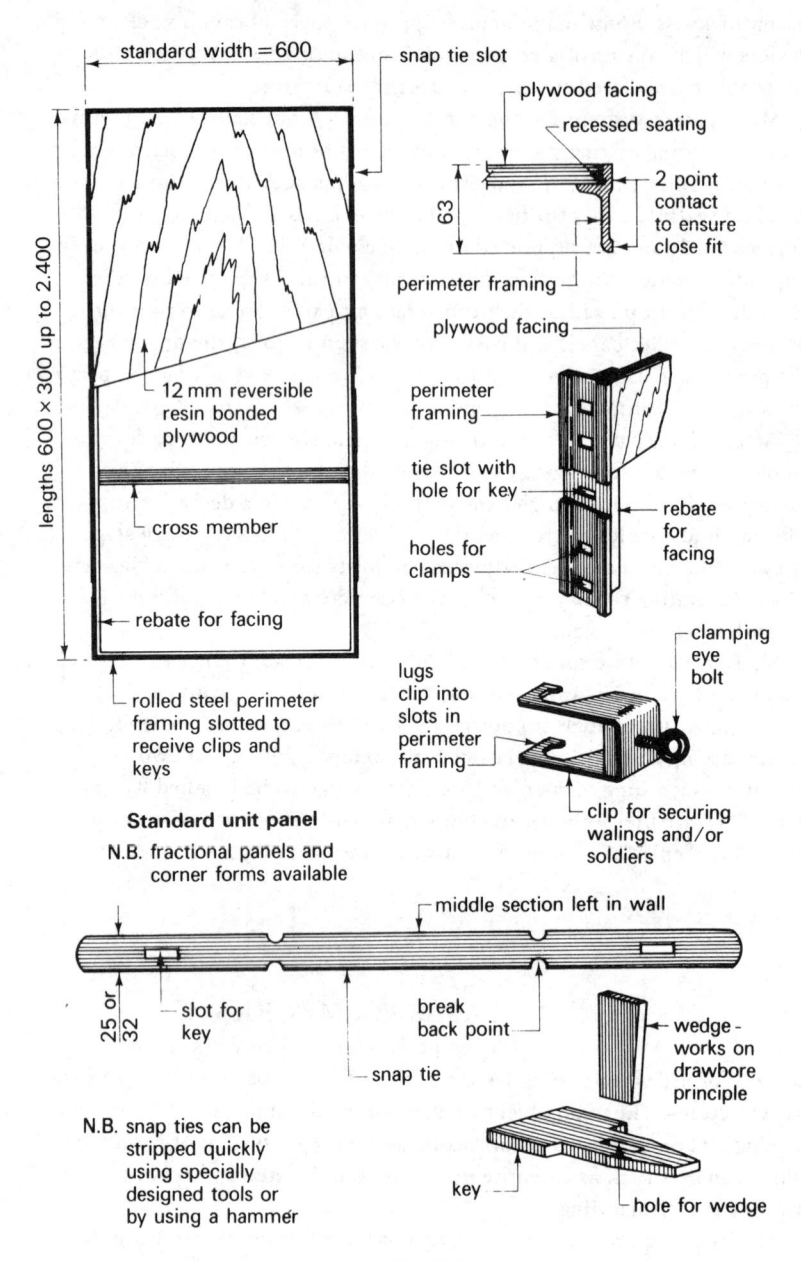

Fig VII.6 System formwork - typical components

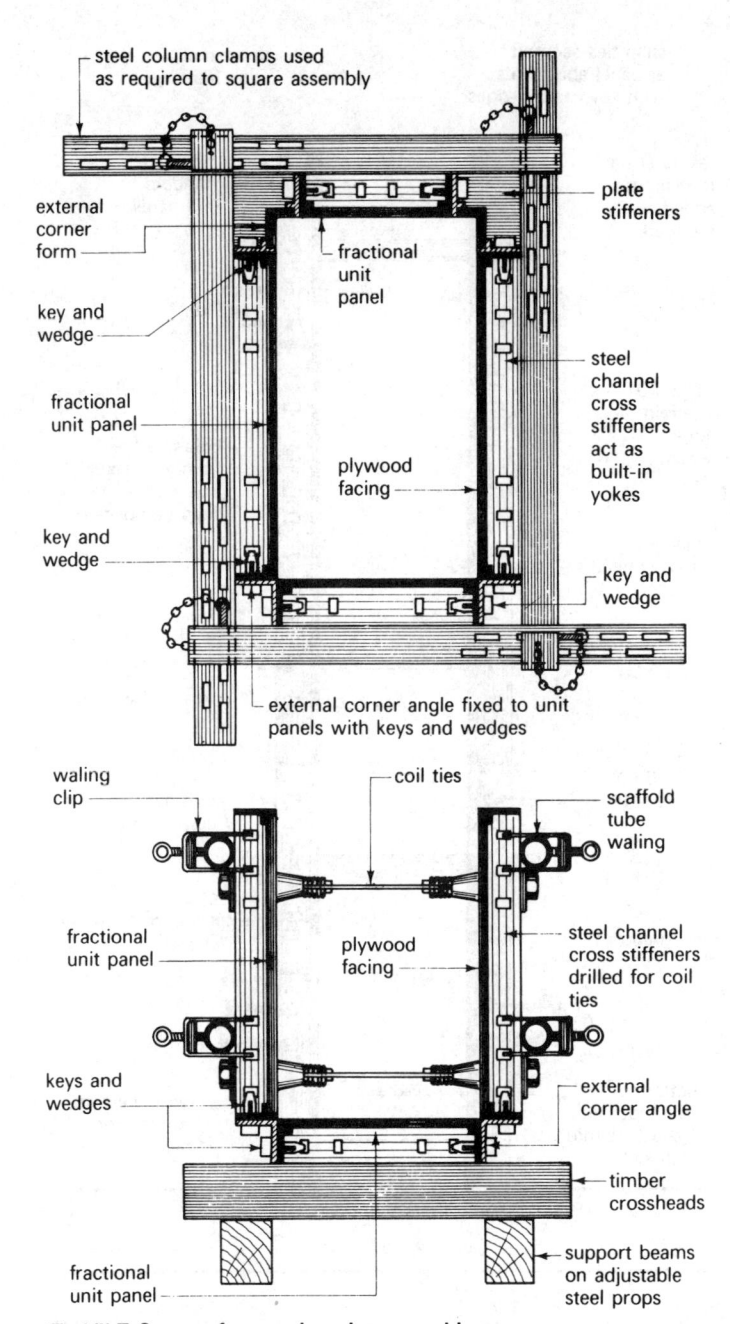

Fig VII.7 System formwork - columns and beams

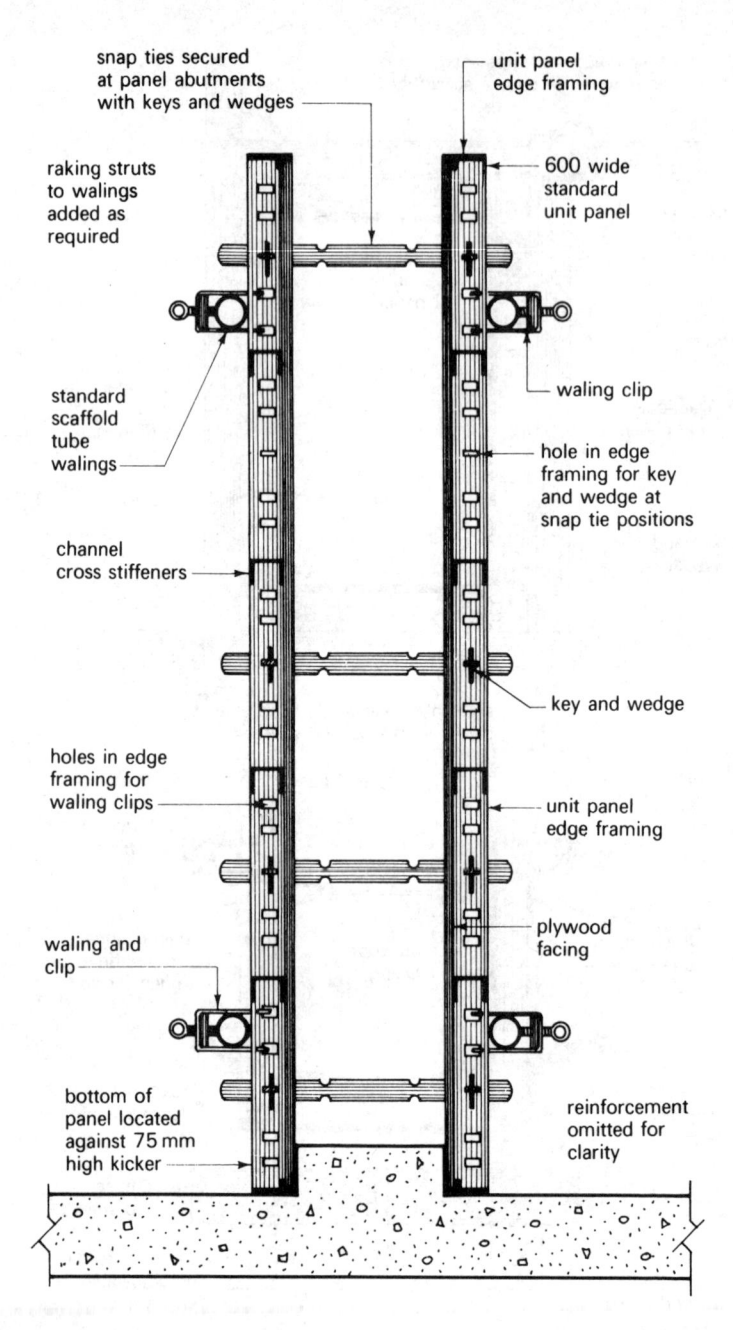

Fig VII.8 System formwork - walls

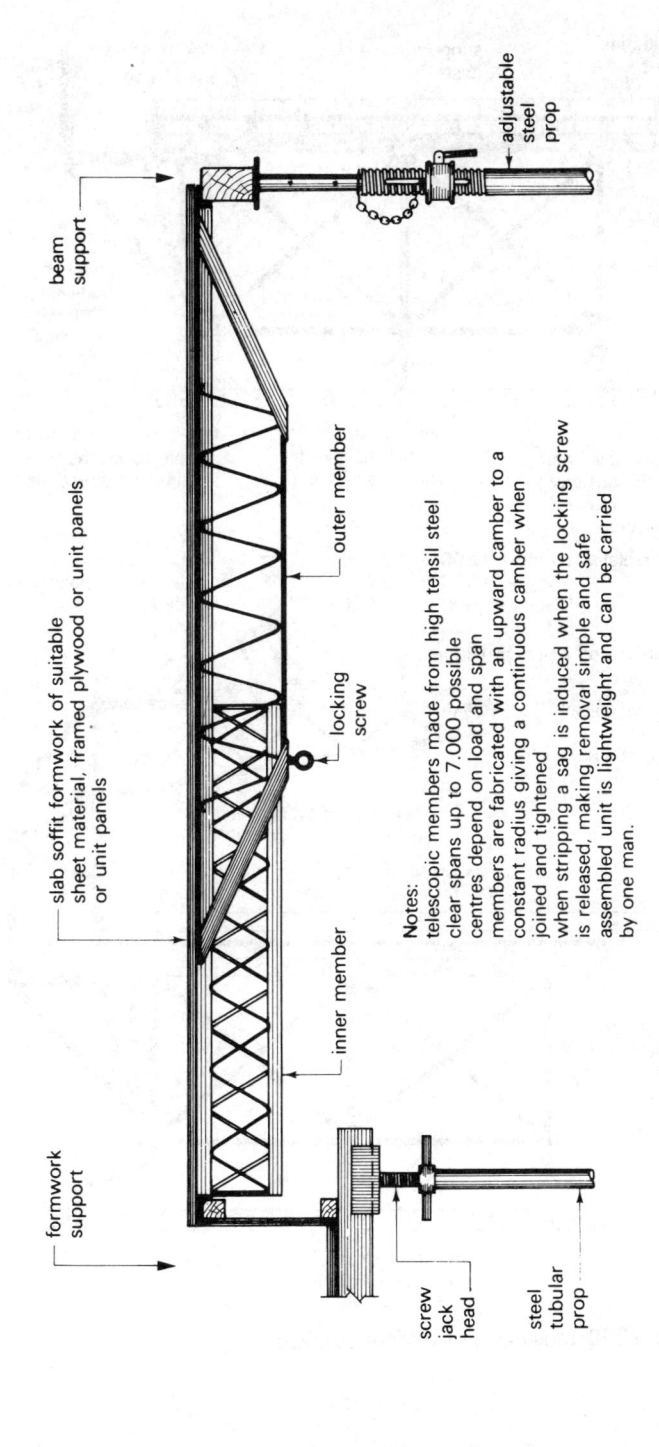

Fig VII.9 System formwork - slabs using telescopic centres

Labels:
- adjustable steel prop
- beam support
- slab soffit formwork of suitable sheet material, framed plywood or unit panels or unit panels
- outer member
- locking screw
- inner member
- formwork support
- screw jack head
- steel tubular prop

Notes:
telescopic members made from high tensile steel clear spans up to 7.000 possible centres depend on load and span members are fabricated with an upward camber to a constant radius giving a continuous camber when joined and tightened when stripping a sag is induced when the locking screw is released, making removal simple and safe assembled unit is lightweight and can be carried by one man.

205

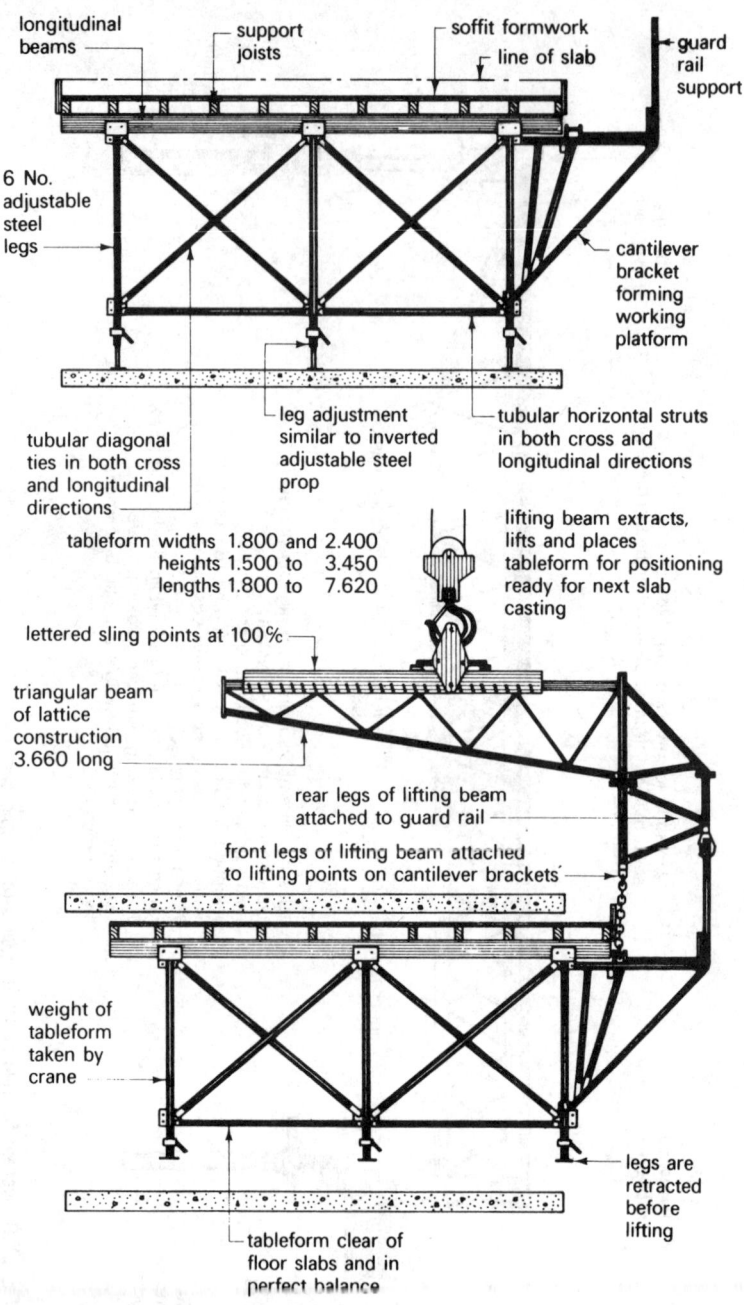

Fig VII.10 Tableforms - 'Kwikform' system

2. Adequate means of lowering the forms so that they can be dropped clear of the newly cast slab; generally the provision for lowering the forms can also be used for final levelling purposes.
3. Means of manoeuvring the forms clear of the structure to a point where they can be attached to the crane for final extraction, lifting and repositioning ready to receive the next concrete pour operation.
4. A means of providing a working platform at the external edge of the slab to eliminate the need for an independent scaffold which would be obstructive to the system.

The basic support members are usually a modified version of inverted adjustable steel props. These props, suitably braced and strutted, carry a framed decking which acts as the soffit formwork. To manoeuvre the forms into a position for attachment to the crane a framed wheeled arrangement can be fixed to the rear end of the tableform so that the whole unit can be moved forward with ease. The tableform is picked up by the crane at its centre of gravity by removing a loose centre board to expose the framework. The unit is then extracted clear of the structure, hoisted in the balanced horizontal position and lowered on to the recently cast slab for repositioning.

Another method, devised by Kwikform Ltd, uses a special lifting beam which is suspended from the crane at predetermined sling points which are lettered so that the correct balance for any particular assembly can be quickly identified, each table having been marked with the letter point required. The lifting beam is connected to the working platform attachment of the tableform so that when the unit formwork is lowered and then extracted by the crane the tableform and the lifting beam are in perfect balance. This method of system formwork is diagrammatically illustrated in Fig. VII.10.

Students should appreciate that the selection of formwork for multistorey and complex structures is not a simple matter, but one which requires knowledge and experience in design appreciation, suitability of materials and site operations. Most of the large building contracting organisations have specialist staff for designing and detailing formwork. These contractors also have experienced and specialist site staff for the fabrication of formwork or if using a patent system to supervise the semi-skilled or unskilled labour being used. Badly designed and/or erected formwork can result in failure of the structure during the construction period or inaccurate and unacceptable members being cast, both of which can be financially disastrous for a company and ruinous to the firm's reputation.

22

Concrete surface finishes

The appearance of concrete members is governed by their surface finish and this is influenced by three major factors namely colour, texture and surface profile. The method used to produce the concrete member will have some degree of influence over the finish obtained since a greater control of quality is usually possible with precast concrete techniques under factory controlled conditions. Most precast concrete products can be cast in the horizontal position which again promotes better control over the resultant casting, whereas *in situ* casting of concrete walls and columns must be carried out using vertical casting techniques which not only has a lesser degree of control but also limits the types of surface treatments which can be successfully obtained direct from the mould or formwork.

COLOUR

The colour of concrete as produced direct from the mould or formwork depends upon the colour of the cement being used and to a lesser extent upon the colour of the fine aggregate. The usual methods of obtaining variations in the colour of finished concrete are:

1. *Using a coloured cement* — the range of colours available is limited and most are pastel shades; if a pigment is used to colour cement the cement content of the mix will need to be increased by approximately 10% to counteract the loss of strength due to the colouring additive.

2. *Using the colour of the coarse aggregate* — the outer matrix or cement paste is removed to expose the aggregate which not only imparts colour to the concrete surface but also texture.

Concrete can become stained during the construction period resulting in a mottled appearance. Some of the causes of this form of staining can be listed as follows:

1. *Using different quality timber within the same form* — generally timber which has a high absorption factor will give a darker concrete than timber with a low absorption factor. The same disfigurement can result from using old and new timber in the same form since older timber tends to give darker concrete than new timber with its probable higher moisture content and hence lower absorption.
2. *Formwork detaching itself from the concrete* — this allows dirt and dust to enter the space and attach itself to the green concrete surface.
3. *Type of release agent used* — generally the thinner the release agent used the better will be the result; even coating over the entire contact surface of the formwork or mould is also of great importance.

Staining occurring on mature concrete, usually after completion and occupation of the building, is very often due to bad design, poor selection of materials or poor workmanship. Large overhangs to parapets and sills without adequate throating can create damp areas which are vulnerable to algae or similar growths and pollution attack. Poor detailing of damp-proof courses can create unsightly stains by failing to fulfil their primary function of providing a barrier to dampness infiltration. Efflorescence can occur on concrete surfaces, although it is not so common as on brick walls. The major causes of efflorescence on concrete are allowing water to be trapped for long periods between the cast concrete and the formwork and poorly formed construction or similar joints allowing water to enter the concrete structure. Removal of efflorescence is not easy or always successful and therefore the emphasis should always be on good design and workmanship in the first instance. Methods of efflorescence removal range from wire brushing, various chemical applications to mechanical methods such as grit-blasting.

TEXTURE

When concrete first became acceptable as a substitute for natural stone in major building works the tendency was to try to recreate the smooth surface and uniform colour possible with natural stones. This kind of finish is difficult to achieve using the medium of concrete for the following reasons:

1. Natural shrinkage of concrete can cause hair line cracks on the surface.
2. Texture and colour can be affected by the colour of the cement, water content, degree of compaction and the quality of the formwork.
3. Pin holes on the surface can be caused by air being trapped between the concrete and the formwork.
4. Rough patches can result from the formwork adhering to the concrete face.
5. Grout leakage from the formwork can cause fins or honeycombing on the cast concrete.

Under site conditions it is not an easy task to keep sufficient control over the casting of concrete members to guarantee that these faults will not occur. A greater control is however possible with the factory type conditions prevailing in a well-organised precast concrete works. It should be noted that attempts to patch, mask or make good the above defects are nearly always visible.

Methods which can be used to improve the appearance of a concrete surface can be listed as follows:

1. Finishes obtained direct from the formwork or mould intended to conceal the natural defects by attracting the eye to a more obvious and visual point.
2. Special linings placed within the formwork to produce a smooth or profiled surface.
3. Removal of the surface matrix to expose the aggregate.
4. Applied surface finishes such as ceramic tiles, renderings and paints.

Formwork can be designed and constructed to highlight certain features such as the joints between form members or between concrete pours or lifts by adding to the inside of the form small fillets to form recessed joints, or conversely by recessing the formwork to form raised joints; the axiom being if you cannot hide or mask a joint make a feature of it. The use of sawn boards, to imprint the grain pattern, can give a visually pleasing effect to concrete surfaces particularly if boards of a narrow width are used.

A wide variety of textured, patterned and profiled surfaces can be obtained by using different linings within the formwork. Typical materials used are thermo-plastics, glass fibre mouldings, moulded rubber and PVC sheets; all of these materials can be obtained to form many various patterns giving many uses and are easily removed from the concrete surface. Glass fibre and thermo-plastic linings have the advantage of being capable of being moulded to any reasonable shape or profile. Ribbed and similar profiles can be produced by fixing materials such as troughed or corrugated steel into the form as a complete lining or as a partial lining to give a panelled effect. Coarse fabrics such as hessian can also be used to produce various textures providing the fabric is first soaked with a light mineral oil and squeezed to remove the excess oil before fixing as a precaution against the fabric becoming permanently embedded on the concrete surface.

The most common method of imparting colour and texture to a concrete surface is to expose the coarse aggregate which can be carried out by a variety of methods depending generally upon whether the concrete is green, mature, or yet to be cast. Brushing and washing is a common method employed on green concrete and is carried out using stiff wire or bristle brushes to loosen the surface matrix and clean water to remove the matrix and clean the exposed aggregate. The process should be carried out as soon as practicable after casting, which is usually within 2 to 6 hours of the initial pour but certainly not more than 18 hours after casting. When treating horizontal surfaces brushing should commence at the edges, whereas brushing to vertical surfaces should start at the base and be finished by washing from the top downwards. Skill is required with this method of exposing aggregates to ensure that only the correct amount of matrix is removed, the depth being dependent on the size of aggregate used. The use of retarding agents can be employed but it is essential that these are only used in strict accordance with the manufacturer's instructions.

Treatments which can be given to mature concrete to expose the aggregate include tooling, bush hammering and blasting techniques. These methods are generally employed when the depth of the matrix to be removed is greater than that usually associated with the brushing method described above. The most suitable age for treating the mature concrete surface depends upon such factors as the type of cement used and the conditions under which the concrete is cured, the usual period being from 2 to 8 weeks after casting.

The usual hand tooling methods which can be applied to natural stone such as point tooling and chiselling can be applied to a mature concrete surface but can be expensive if used for large areas. Tooling methods should not be used on concrete if gravel aggregates have been specified since these tend to shatter and leave unwanted pits on the surface. Hard point tooling near to sharp arrises should also be avoided because of the tendency to spalling at these edges; to overcome this problem a small plain margin at least 10 mm wide should be specified.

Except for small areas or where special treatment is required bush hammering has largely replaced the hand tooling techniques to expose the aggregate. Bush hammers are power-operated, hand-held tools to which two types of head can be used to give a series of light but rapid blows (e.g. 1 750 blows per minute) to remove about 3 mm of matrix for each passing. Circular heads with 21 cutting points are easier to control if working on confined or small areas than the more common and quicker roller head with its 90 cutting points.

Blasting techniques using sand, shot or grit are becoming increasingly popular since these methods do not cause spalling at the edges and generally result in a very uniform textured surface. Blasting can be carried out at almost any time after casting but preferably within 16 hours to 3 days of casting which gives a certain degree of flexibility to a programme of work. This method is of a specialist nature and is normally carried out by a sub-contractor who can usually expose between 4 to 6 m^2 of surface per hour. Sand is usually dispersed in a jet of water whereas grit is directed on to the concrete surface from a range of about 300 mm in a jet of compressed air and is the usual method employed. Shot is dispersed from a special tool with an enclosed head which collects and recirculates the shot.

Any method of removing the surface skin or matrix will reduce the amount of concrete cover over the reinforcement and therefore if this type of surface treatment is to be used an adequate cover of concrete should be specified such as 45 mm minimum before surface treatment.

When exposing aggregates by the above methods the finished result cannot be assured since the distribution of the aggregates throughout the concrete is determined by the way in which the fine and coarse aggregates disperse themselves during the mixing, placing and compacting operations. If a particular colour or type of exposed aggregate is required it will be necessary to use the chosen aggregate throughout the mix which may be both uneconomic and undesirable. A method which can be used for casting *in situ* concrete which will ensure the required distribution of aggregate over the surface and will enable a selected aggregate to be used on the surface is the aggregate transfer method.

This method entails sticking the aggregate to the rough side of pegboard sheets with a mixture of water-soluble cellulose compounds and sand fillers. The resultant mixture has a cream-like texture and is spread evenly over the pegboard surface to a depth of one third the aggregate size. The aggregate may be sprinkled over the surface and lightly tamped or alternatively placed by hand after which the prepared board is allowed to set and dry for about 36 hours. This prepared pegboard is used as a liner to the formwork with a loose baffle of hardboard or plywood immediately in front of the aggregate face as protection during the pouring of the cement. The baffle is removed as work proceeds. This is not an easy method of obtaining a specific exposed aggregate finish to *in situ* work and it is generally recommended that where possible simpler and just as effective precast methods should be considered.

Precast concrete casting can be carried out in horizontal or vertical moulds. Horizontal casting can have surface treatments carried out on the upper surface or the lower surface which is in contact with the mould base. Plain smooth surfaces can be produced by hand trowelling the upper surface but this is expensive in labour costs and requires a high degree of skill if a level surface of uniform texture and colour is to be obtained, whereas a good plain surface can usually be achieved direct from the mould using the lower face method. Various surface textures can be obtained from upper face casting by using patterned rollers, profiled tamping boards, using a straight edged tamping board drawn over the surface with a sawing action and by scoring the surface with brooms or rakes. Profiled finishes from lower face casting can be obtained by using a profiled mould base or a suitable base liner without any difficulty.

Exposed aggregate finishes can be obtained from either upper or lower face horizontal casting by spraying and brushing or by the tooling methods previously described. Alternatives for upper face casting are to trowel into the freshly cast surface selected aggregates or to trowel in a fine (10 mm) aggregate dry mortar mix and using a float with a felt pad to pick up the fine particles and cement leaving a clean exposed aggregate finish.

The sand bed method of covering the bottom of the mould with a layer of sand and placing the selected aggregate into the sand bed and immediately casting the concrete over the layer of aggregate is the usual method employed with lower face casting. Cast on finishes such as bricks, tiles and mosaic can be fixed in a similar manner to those described above for exposed aggregates. For all practical purposes vertical casting of precast concrete members needs the same considerations as those described previously in the context of *in situ* casting.

Another surface treatment which could be considered for concrete is grinding and polishing although this treatment is expensive and time consuming, being usually applied to surfaces such as slate, marble and terrazzo. The finish can be obtained by using a carborundum rotary grinder incorporating a piped water supply to the centre of the stone. The grinding operation should be carried out as soon as the mould or formwork has been removed by first wetting the surface and then grinding, allowing the stone dust and water to mix forming an abrasive paste. The operation is completed by washing and brushing to remove the paste residue. Dry grinding is possible but this creates a great deal of dust from which the operative's eyes and lungs must be protected by wearing protective masks.

It is very often wrongly assumed that by adding texture or shape to a concrete surface any deficiency in design, formwork fabrication or workmanship can be masked. This is not so; a textured surface, like the plain surface, will only be as good as the design and workmanship involved in producing the finished component and any defects will be evident with all finishes.

Part VIII
Stairs

23 Concrete stairs

Any form of stairs is primarily a means of providing circulation and communication between the various levels within a building. Apart from this primary function stairs may also be classified as a means of escape in case of fire; if this is the case the designer is severely limited by necessary regulations as to choice of materials, position and sizing of the complete stairway. Stairs which do not fulfil this means of escape function are usually called accommodation stairs and as such are not restricted by the limitations given above for escape stairs.

Escape stairs have been covered in a previous section of this volume and it is only necessary to reiterate the main points:

1. Constructed from non-combustible materials.
2. Stairway protected by a fire-resisting enclosure.
3. Separated from the main floor area by a set or sets of self-closing fire-resisting doors.
4. Limitations as to riser heights, tread lengths and handrail requirement usually based upon use of building.
5. In common with all forms of stairs, all riser heights must be equal throughout the rise of the stairs.

It should be appreciated that points 2 and 3 listed above can prove to be inconvenient to persons using the stairway for general circulation within the building, such as having to pass through self-closing doors, but in the context of providing a safe escape route this is unavoidable.

Generally concrete escape stairs are designed in straight flights with not more than 16 risers per flight providing such flights are not constructed in line. A turn of at least 30° is required after any flight which has more than thirty-six risers. In special circumstances such as low usage (not more than 50 persons), minimum overall diameter of 1.500 m and a total rise not exceeding 3.000 m spiral stairs may be acceptable as a suitable means of escape in case of fire.

The construction of simple reinforced concrete stairs together with a suitable formwork arrangement is usually covered in the second year of a course in construction technology (see Volume, 2 Part V) and it is a useful exercise, at this stage, for the student to refresh his memory on the following basic requirements:

1. Concrete mix usually specified as 1:2:4/20 mm aggregate.
2. Minimum cover of concrete over reinforcement 15 mm or bar diameter whichever is the greater to give a 1-hour fire resistance.
3. Waist thickness usually between 100 and 250 mm depending on stair type.
4. Mild steel or high yield steel bars can be used as reinforcement.
5. Continuous handrails of non-combustible materials at a height of 900 mm above the pitch line are required to all stairs and to both sides if the stair width exceeds 1.000 m.

The main area of study for an advanced level course is concerned with the different stair types or arrangements which can range from a single straight flight to an open spiral stair.

Single straight flight stair: this form of stair, although simple in design and construction, is not popular because of the plan space it occupies. In this stair arrangement the flight behaves as a simple supported slab spanning from landing to landing. The effective span or total horizontal going is usually taken as being from landing edge to edge by providing a downstand edge beam to each landing. If these edge beams are not provided the effective span would be taken as overall of the landings, resulting in a considerably increased bending moment and hence more reinforcement. Typical details are shown in Fig. VIII.1.

Inclined slab stair with half space landings: these stairs have the usual plan format for reinforced concrete stairs giving a more compact plan layout and better circulation than the single straight flight stairs. The half space or 180° turn landing is usually introduced at the mid-point of the rise giving equal flight spans, thus reducing the effective span and hence the bending moment considerably. In most designs the landings span crosswise on to a load bearing wall or beam and the flights span from landing to landing. The point of intersection of the soffits to the flights with the landing soffits can be detailed in one of two ways:

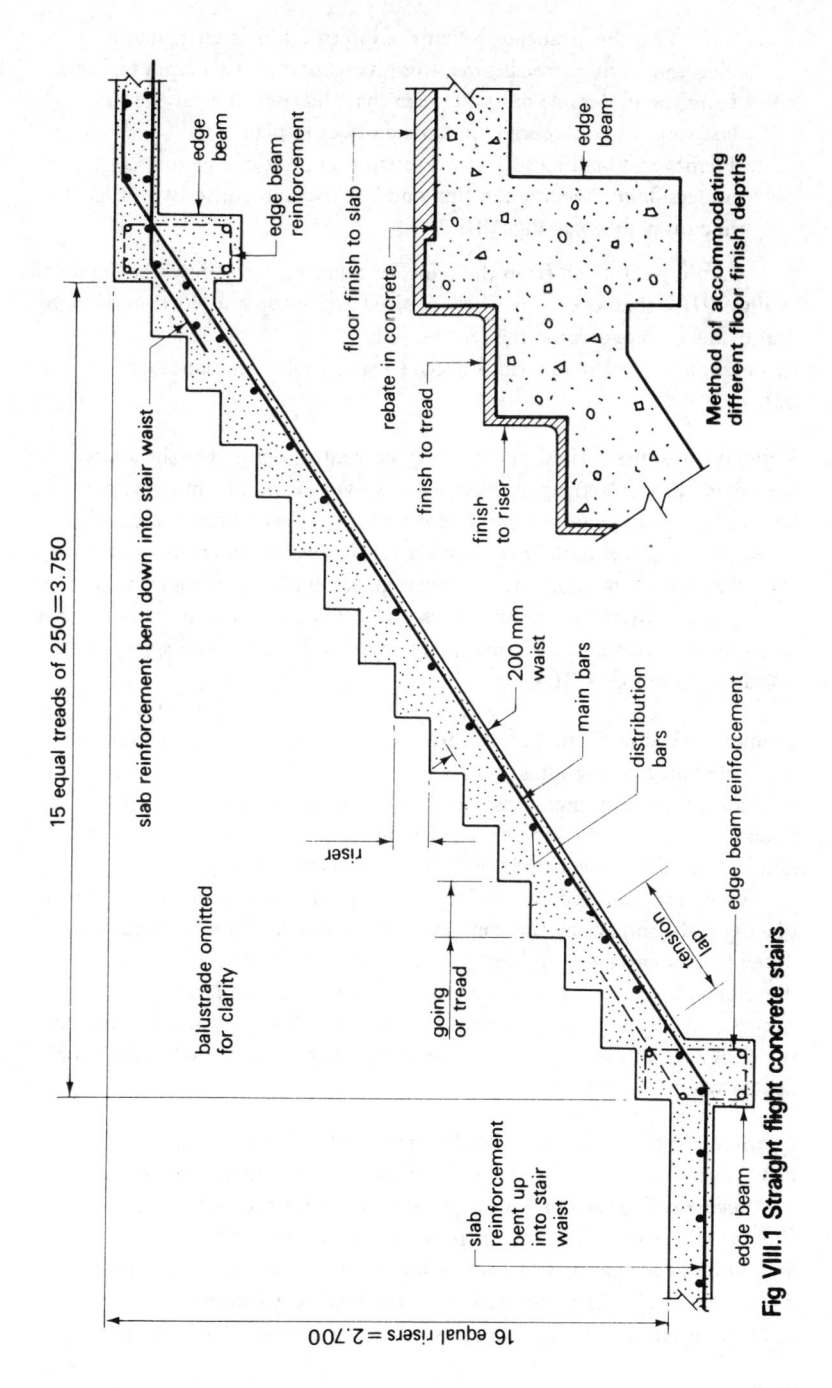

Fig VIII.1 Straight flight concrete stairs

1. Soffits can be arranged so that the intersection or change of direction is in a straight line; this gives a better visual appearance from the underside but will mean that the riser lines of the first and last steps in consecutive flights are offset in plan.
2. Flights and landing soffit intersections are out of line on the underside by keeping the first and last risers in consecutive flights in line on plan — see Fig. VIII.2.

It should be noticed from the reinforcement pattern shown in the detail in Fig. VIII.2 that a tension lap is required at the top and bottom of each flight, this is to overcome the tension induced by the tendency of the external angles of the junctions between stair flights and landings to open out.

String beam stairs: these stairs are an alternative design for the stairs described above. A string or edge beam is used to span from landing to landing to resist the bending moment with the steps spanning crosswise between them; this usually results in a thinner waist dimension and an overall saving in the concrete volume required but this saving in material is usually offset by the extra formwork costs. The string beams can be either upstand or downstand in format and to both sides if the stairs are free standing — see Fig. VIII.3.

Cranked slab stairs: these stairs are very often used as a special feature since the half space landing has no visible support being designed as a cantilever slab. Bending, buckling and torsion stresses are induced with this form of design creating the need for reinforcement to both faces of the landing and slab or waist of the flights; indeed the amount of reinforcement required can sometimes create site problems with regard to placing and compacting the concrete. The problem of deciding upon the detail for the intersection line between flight and landing soffits, as described above for inclined slab stairs with half spaced landings, also occurs with this stair arrangement. Typical details of a cranked slab stair, which is also known as a continuous stair, scissor stair or jack knife stair, are shown in Fig. VIII.4.

Cantilever stairs: sometimes called spine wall stairs and consists of a central vertical wall from which the flights and half space landings are cantilevered. The wall provides a degree of fire resistance between the flights and are therefore used mainly for escape stairs. Since both flights and landings are cantilevers the reinforcement is placed in the top of the flight slab and in the upper surface of the landing to counteract the induced negative bending moments. The plan arrangement can be a single

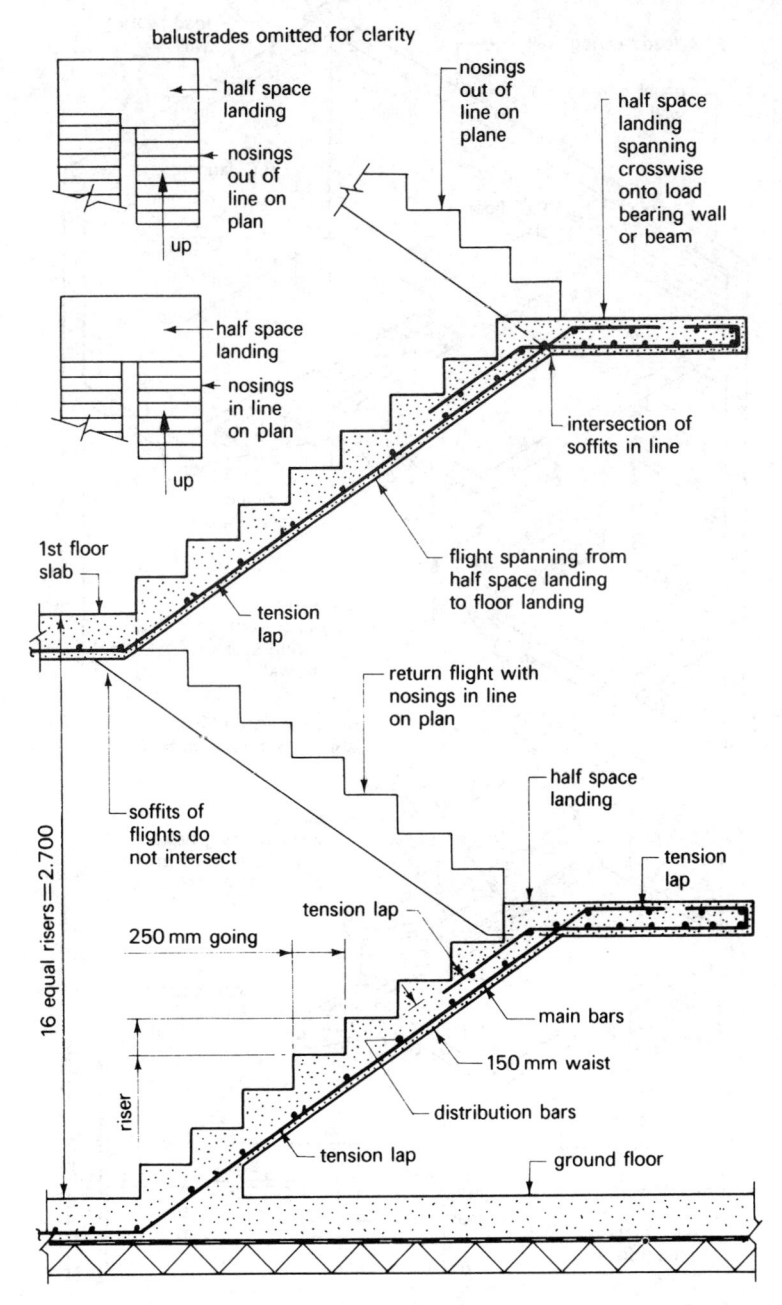

Fig VIII.2 Inclined slab concrete stair with half space landings

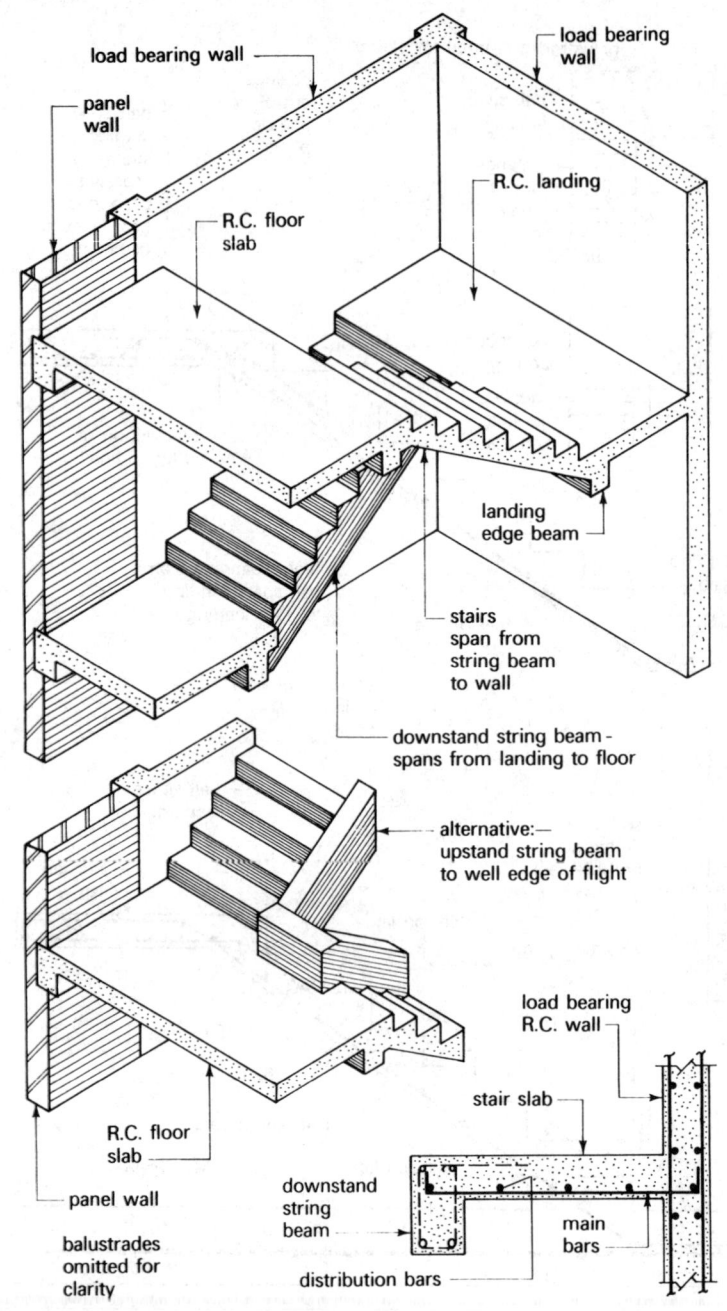

Fig VIII.3 String beam concrete stairs

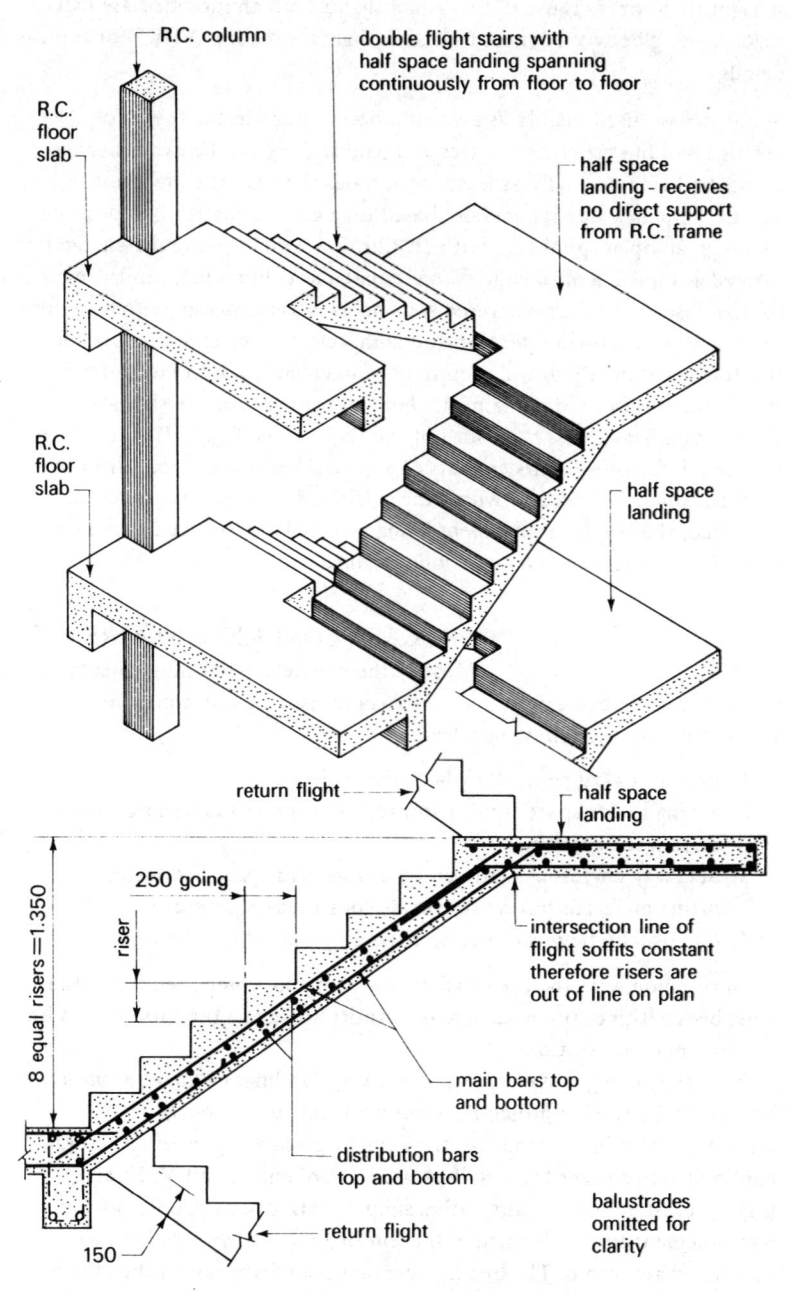

Fig VIII.4 Cranked or continuous slab concrete stairs

221

straight flight or, as is usual, two equal flights with an intermediate half space landing between consecutive stair flights — see Fig. VIII.5 for typical details.

Spiral stairs: used mainly as accommodation stairs in the foyers of prestige buildings such as theatres and banks. They can be expensive to construct being normally at least seven times the cost of conventional stairs. The plan shape is generally based on a circle although it is possible to design an open spiral stair with an elliptical core. The spiral stair can be formed around a central large diameter circular column in a similar manner to that described for cantilevered stair or, as is usual, design with a circular open stair well. Torsion, tension and compressive stresses are induced in this form of stair which will require reinforcement to both faces of the slab in the form of radial main bars bent to the curve of the slab with distribution bars across the width of the flight — see Fig. VIII.6. Formwork for spiral stairs consists of a central vertical core or barrel to form the open stair well to which the soffit and riser boards are set out and fixed, the whole arrangement being propped and strutted as required from the floor level in a conventional manner.

PRECAST CONCRETE STAIRS

Most of the concrete stair arrangements previously described can be produced as precast concrete components which can have the following advantages:

1. Better quality control of the finished product.
2. Saving in site space, since formwork storage and fabrication space is no longer necessary.
3. Stairway enclosing shaft can be utilised as a space for hoisting or lifting materials during the major construction period.
4. Can usually be positioned and fixed by semi-skilled labour.

In common with the use of all precast concrete components the stairs must be repetitive and in sufficient quantity to justify their use and to be an economic proposition.

The straight flight stair spanning between landings can have a simple bearing or, by leaving projecting reinforcement to be grouted into preformed slots in the landings, they can be given a degree of structural continuity; this latter form is illustrated in Volume 2, Part V. Straight flight precast concrete stairs with a simple bearing require only bottom reinforcement to the slab and extra reinforcement to strengthen the bearing rebate or nib. The bearing location is a rebate cast in the *in situ* floor slab or landing leaving a tolerance gap of 8 to 12 mm which is filled

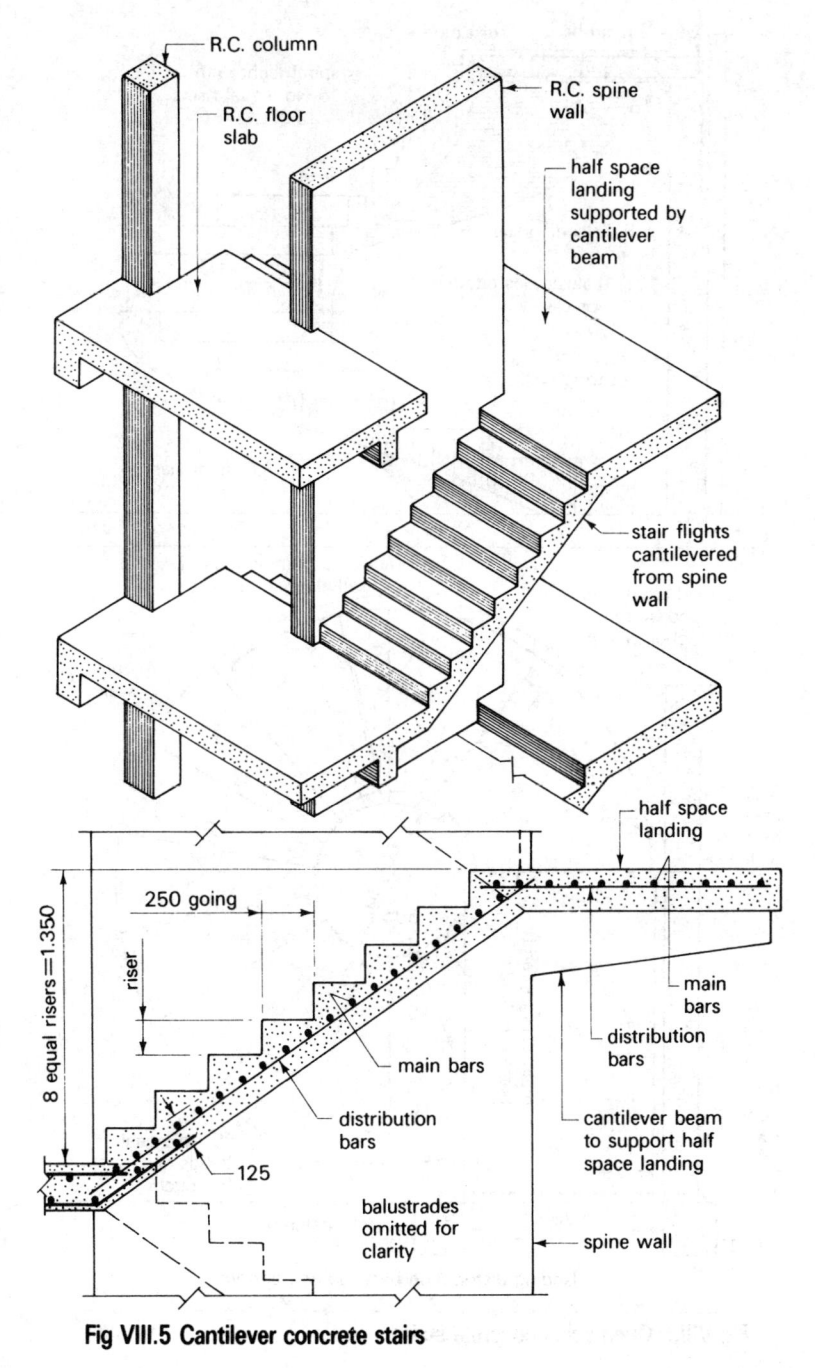

Fig VIII.5 Cantilever concrete stairs

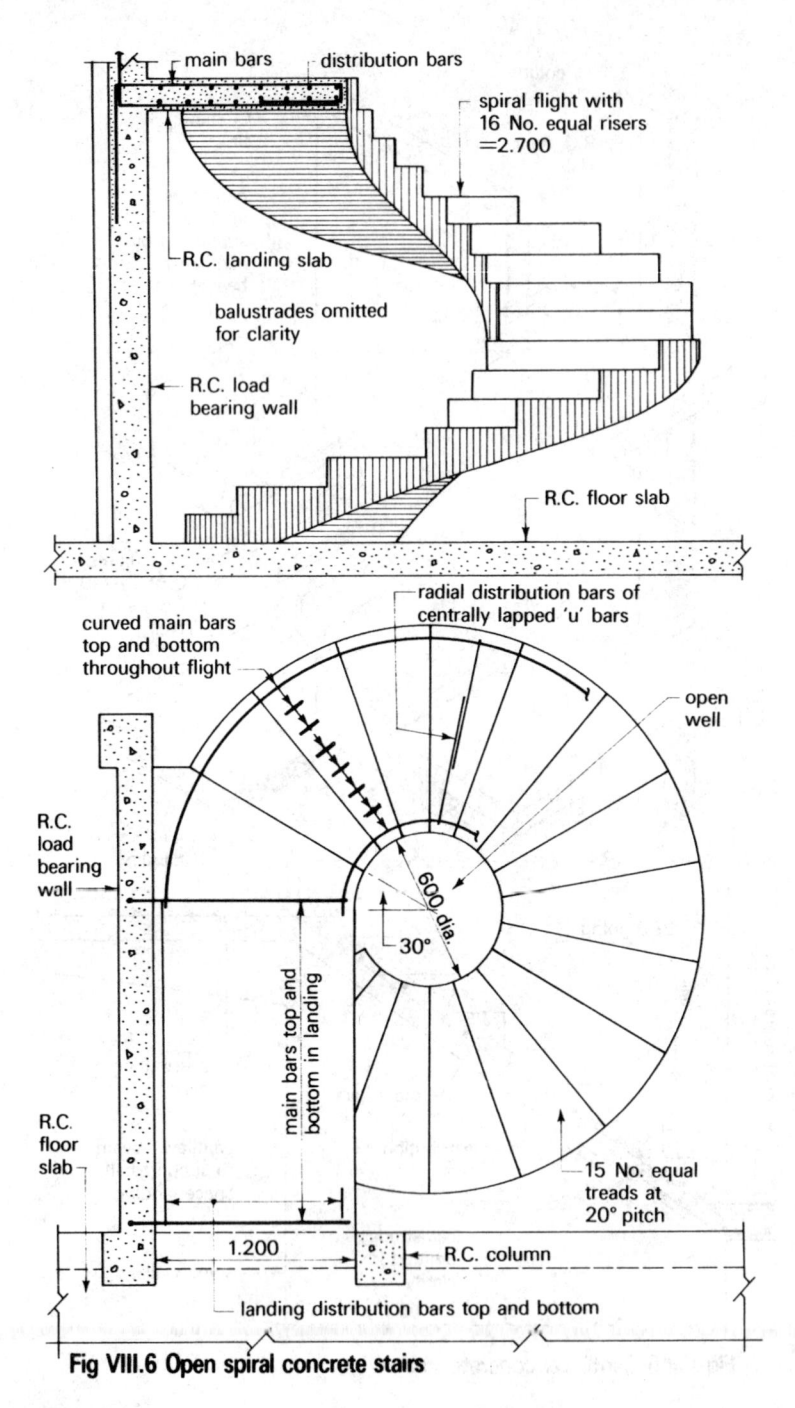

Fig VIII.6 Open spiral concrete stairs

with a compressible material to form a flexible joint. The decision as to whether the stair and landing soffits will be in line or, alternatively, the first and last risers kept in line on plan remains, and since the bearing rebates are invariably cast in a straight line to receive both upper and lower stair flights the intersection design is detailed with the precast units — a typical example is shown in Fig. VIII.7.

Cranked slab precast concrete stairs are usually formed as an open well stair. The bearing for the precast landings to the *in situ* floor or to the structural frame is usually in the form of a simple bearing as described above for straight flight precast concrete stairs. The infill between the two adjacent flights, in an open well plan arrangement at floor and intermediate landing levels, can be of *in situ* concrete with structural continuity provided by leaving reinforcement projecting from the inside edge of the landings — see Fig. VIII.8. It must be remembered that when precast concrete stair flights are hoisted into position different stresses may be induced from those which will be encountered in the fixed position. To overcome this problem the designer can either reinforce the units for both conditions or as is more usual provide definite lifting points in the form of projecting lugs or by utilising any holes cast in to receive the balustrading.

Precast open riser stairs are a form of stair which can be both economic and attractive consisting of a central spine beam in the form of a cut string supporting double cantilever treads of timber or precast concrete. The foot of the lowest spine beam is located and grouted into a preformed pocket cast in the floor whereas the support at landing and floor levels is a simple bearing located in a housing cast into the slab edge — see Fig. VIII.9. Anchor bolts or cement in sockets are cast into the spine beam to provide the fixing for the cantilever treads. The bolt heads are recessed below the upper tread surface and grouted over with a matching cement mortar for precast concrete treads and concealed with matching timber pellets when hardwood treads are used. The supports for the balustrading and handrail are located in the holes formed at the ends of the treads and secured with a nut and washer on the underside of the tread. It will have been noticed that the balustrade and handrail details have been omitted from all the stair details so far considered; this has been done for reasons of clarity and this aspect of stairwork is usually covered during the introductory work in the second year of a typical four year course of study in construction technology — see Volume 2, Part V.

Spiral stairs in precast concrete work are based upon the stone stairs found in many historic buildings such as Norman castles and cathedrals, consisting essentially of steps which have a 'keyhole' plan shape rotating round a central core. Precast concrete spiral stairs are usually open riser

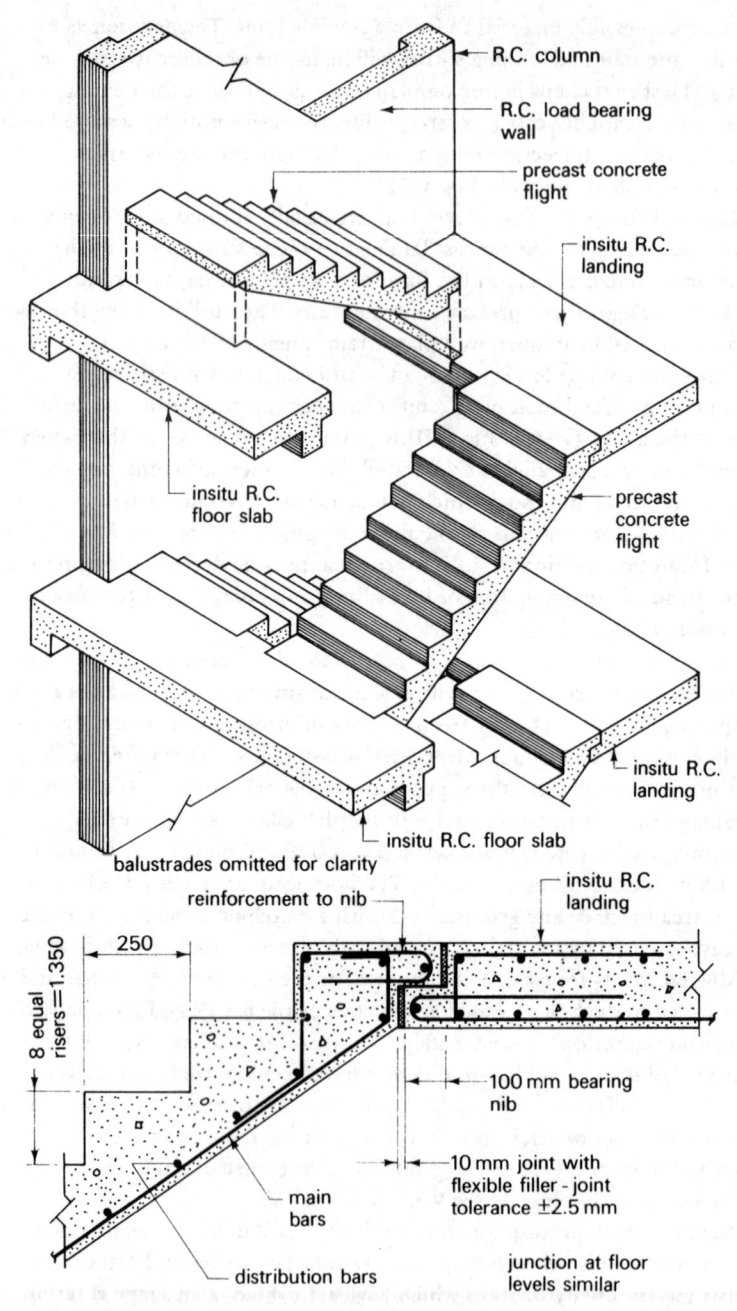

Fig VIII.7 Precast concrete straight flight stairs

226

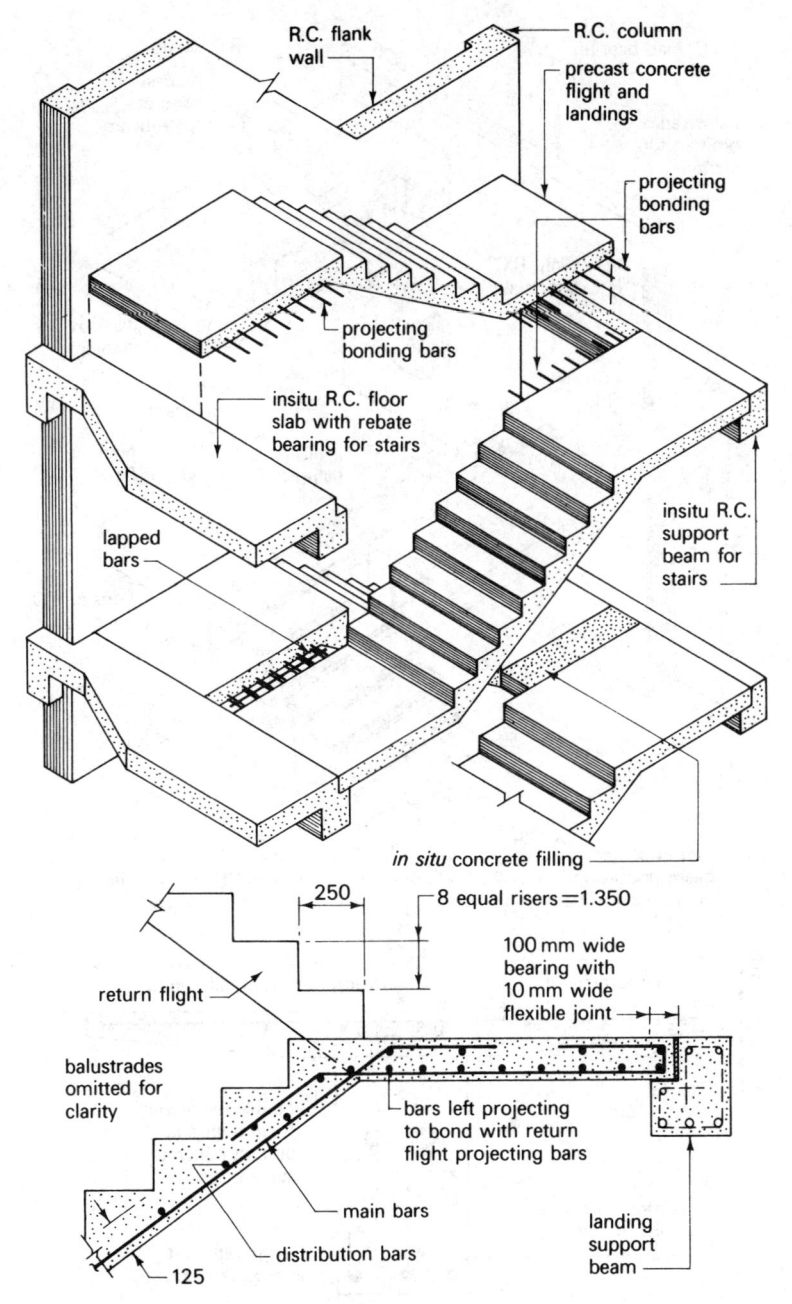

Fig VIII.8 Precast concrete cranked slab stairs

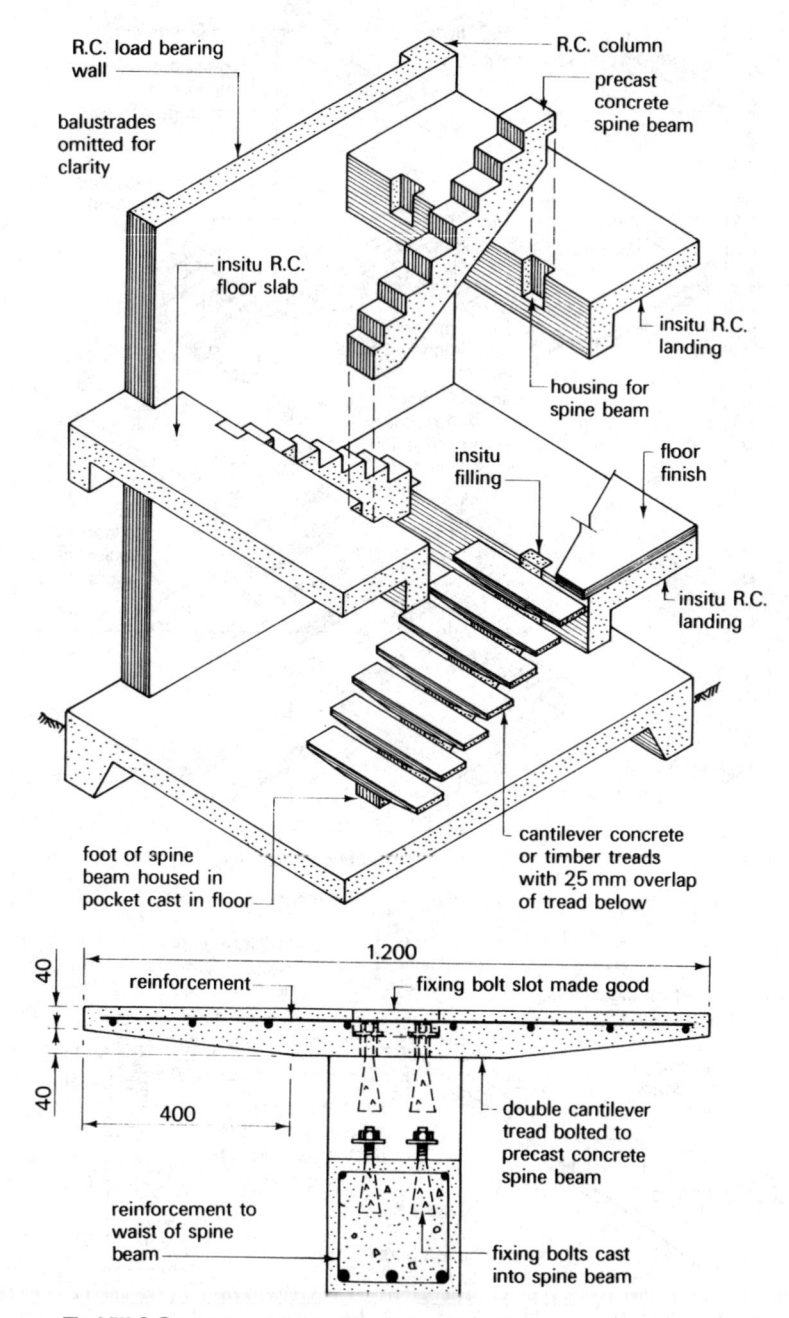

Fig VIII.9 Precast concrete open riser stairs

Fig VIII.10 Precast concrete spiral stairs

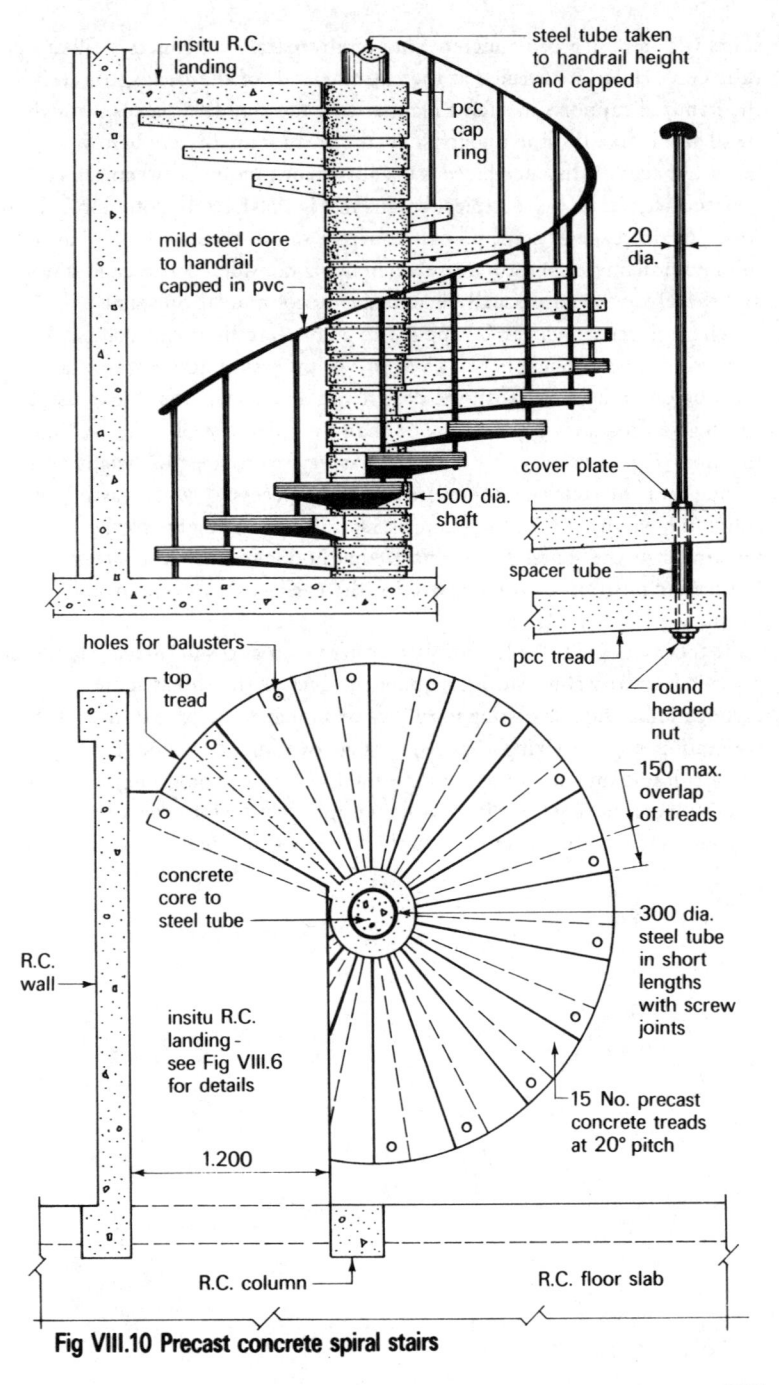

stairs with a reinforced concrete core or alternatively a concrete-filled steel tube core. Holes are formed at the extreme ends of the treads, to receive the handrail supports in such a manner that the standard passes through a tread and is fixed to the underside of the tread immediately below. A hollow spacer or distance piece is usually incorporated between the two consecutive treads — see typical details in Fig. VIII.10. In common with all forms of this type of stair, precast concrete spiral stairs are limited as to minimum diameter and total rise when being considered as escape stairs and therefore they are usually installed as accommodation stairs.

The finishes which can be applied to a concrete floor can also be applied in the same manner to an *in situ* or precast concrete stair. Care must be taken however with the design and detail since the thickness of finish given to stairs is generally less than the thickness of a similar finish to floors. It is necessary, for reasons of safety, to have equal height risers throughout the stair rise; therefore it may be necessary when casting the stairs to have the top and bottom risers of different heights to the remainder of the stairs. An alternative method is to form a rebate at the last nosing position to compensate for the variance in floor and stair finishes — see Fig. VIII.1.

If the stairs are to be left as plain concrete an anti-slip surface should be provided by trowelling into the upper surfaces of the treads some carborundum dust or casting in rubber or similar material grip inserts to the leading edge or fixing a special nosing covering of aluminium alloy or other suitable metal with a grip patterned surface or containing non-slip inserts. Metal nosing coverings with an upper grip surface can also be used in conjunction with all types of applied finishes to stairs.

24
Metal stairs

Metal stairs can be constructed to be used as escape stairs or accommodation stairs both internally and externally. Most metal stairs are manufactured from mild steel with treads of cast iron or mild steel and in straight flights with intermediate half space landings. Spiral stairs in steel are also produced but their use as an escape stair is limited by size and the number of persons likely to use the stairway in the event of a fire. Aluminium alloy stairs are also made and are used almost exclusively as internal accommodation stairs.

Since the layout of most buildings is different stairs are very often purpose made to suit the particular situation. Concrete being a flexible material at the casting stage generally presents little or no problems in this respect, whereas purpose-made metal stairs can be more expensive and take longer to fabricate in the workshop. Metal spiral stairs have the distinct advantage that the need for temporary support and hoisting equipment is eliminated. All steel stairs have the common disadvantage of requiring regular maintenance in the form of painting as a protection against corrosion.

Most metal stairs are supplied in a form which requires some site fabrication and this is usually carried out by the supplier's site erection staff, the main contractor having been supplied with the necessary data as to foundation pads, holding down bolts, any special cast-in fixings and any pockets to be left in the structural members or floor slabs to enable this preparatory work to be completed before the stairs are ready to be fixed.

Steel escape stairs: these have already been considered in the context of means of escape in case of fire and a typical general arrangement is illustrated in Fig. V.27 which shows a structural steel support frame and a stairway composed of steel plate strings with preformed treads giving an open riser format. The treads for this type of stair are bolted to the strings and can be of a variety of types ranging from perforated cast iron to patterned steel treads with renewable non-slip nosings. Handrail balustrades or standards can be of steel square or tubular sections bolted to the upper surface of a channel string or to the side of a channel or steel plate string. Figure VIII.11 shows typical steel escape stair components and should be read in conjunction with the stair arrangement shown in Fig. IV.27.

Steel spiral stairs: these may be allowed as an internal or external means of escape stairs if they are not for more than 50 persons, the maximum total rise is 9 000 and the minimum overall diameter is 1.500 m. Spiral stairs give a very compact arrangement and can be the solution in situations where plan area is limited. In common with all steel external escape stairs the tread and landing plates should have a non-slip surface and be self draining with the stairway circulation width completely clear of any opening doors. Two basic forms are encountered, namely those with treads which project from the central pole or tube and those which have riser legs. The usual plan format is to have 12 or 16 treads to complete one turn around the central core and terminating at floor level with a quarter circle landing or square landing. The standards, like those used for precast concrete spiral stairs, pass through one tread and are secured on the underside of the tread immediately below, giving strength and stability to both handrail and steps. Handrails are continuous and usually convex in cross section of polished metal, painted metal or plastic covered. Typical details are shown in Fig. VIII.12.

String beam steel stairs: used mainly to form accommodation stairs which need to be light and elegant in appearance; this is achieved by using small sections and an open riser format. The strings can be of mild steel tube, steel channel, steel box or small universal beam sections fixed by brackets to the upper floor surfaces or landing edges to act as inclined beams. The treads, which can be of hardwood timber, precast concrete or steel, are supported by plate, angle or tube brackets welded to the top of the string beam. Balustrading can be fixed through the ends of the treads or alternatively supported by brackets attached to the outer face of the string beam — typical details are shown in Fig. VIII.13.

Pressed steel stairs: accommodation stairs made from light pressed metal such as mild steel. Each step is usually pressed as one unit with the tread

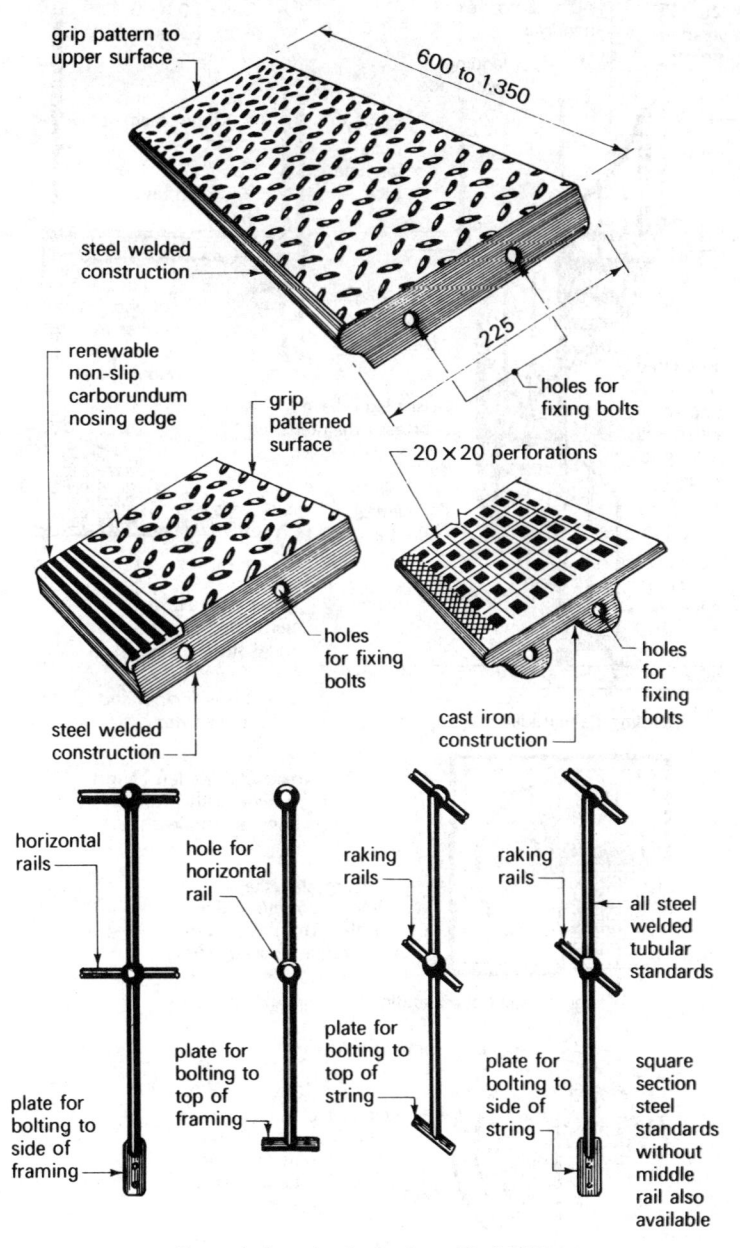

For typical steel stair detail see Fig IV.27

Fig VIII.11 Typical steel stair components

233

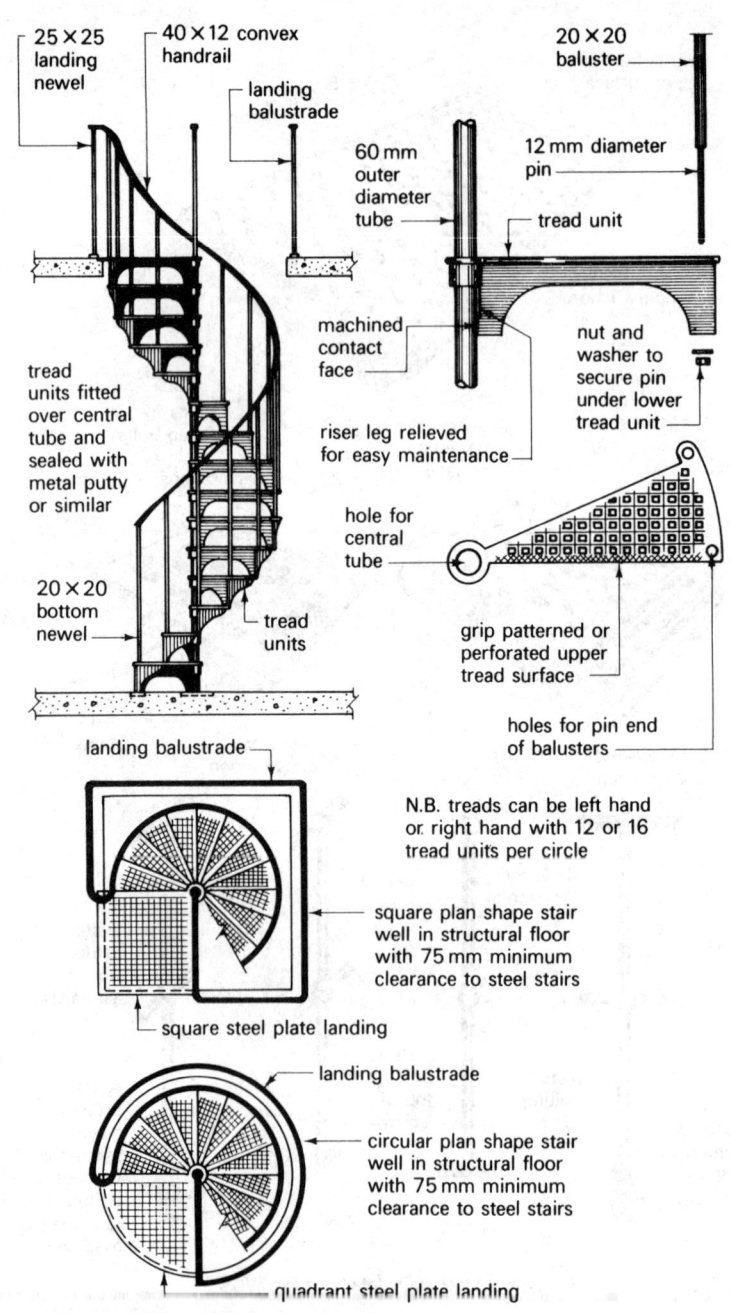

Fig VIII.12 Typical steel spiral stairs

Fig VIII.13 Typical steel string beam stair details

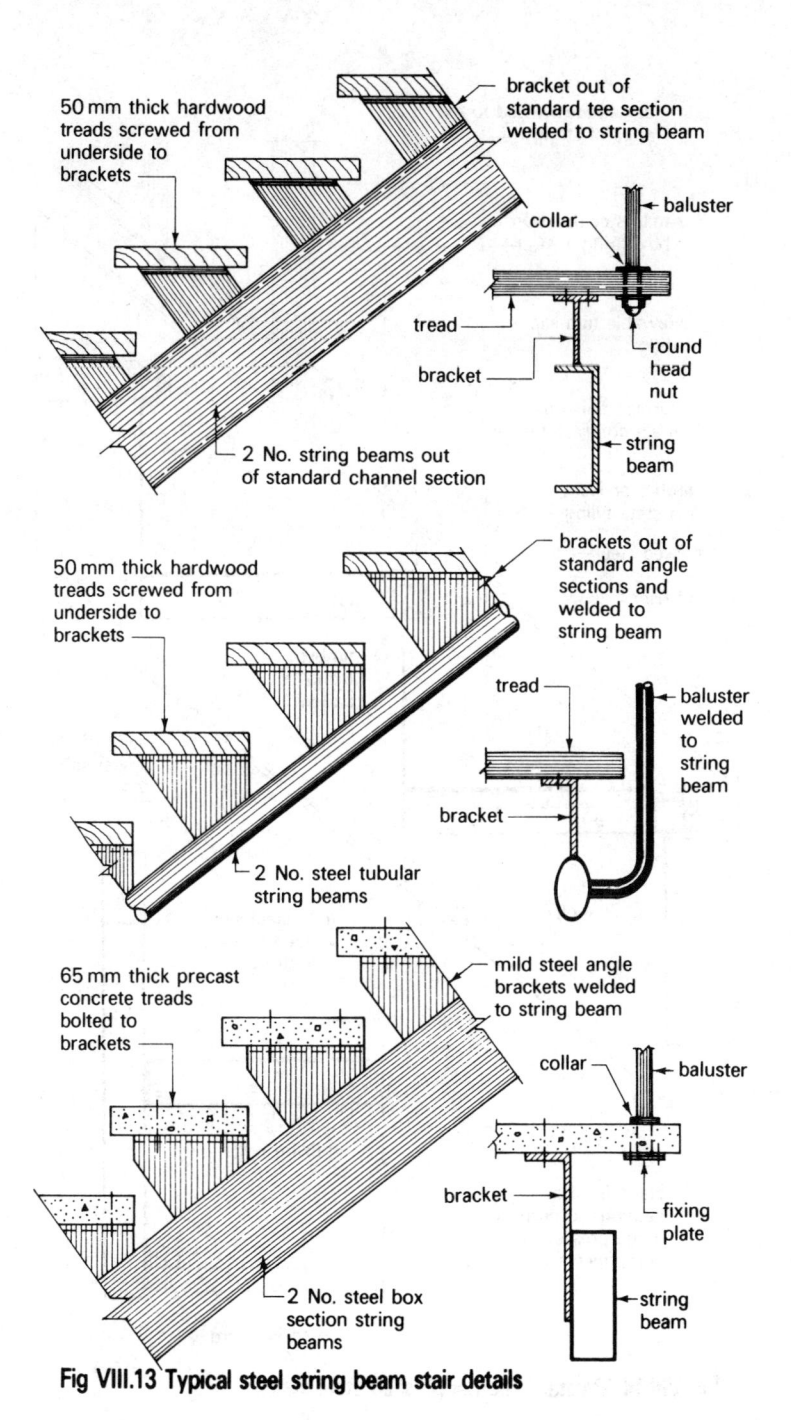

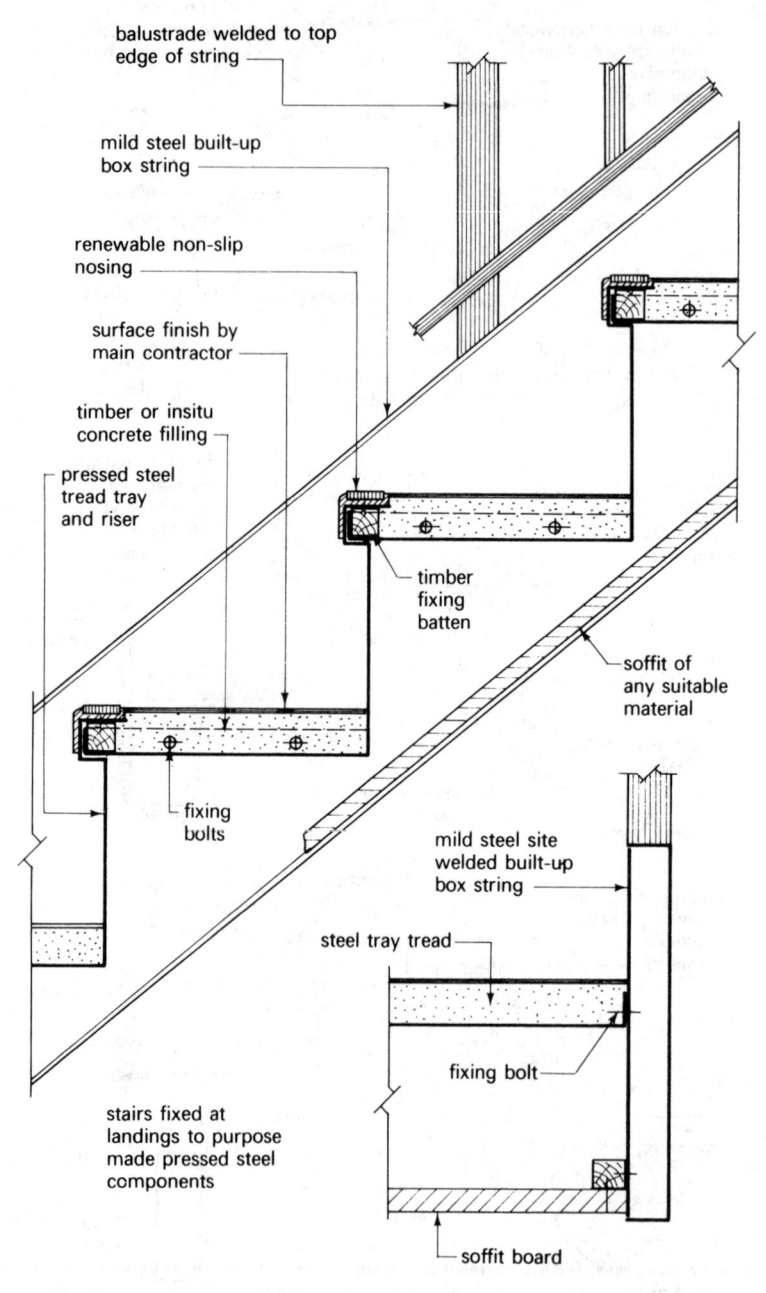

Fig VIII.14 'Prestair' internal pressed steel stairs

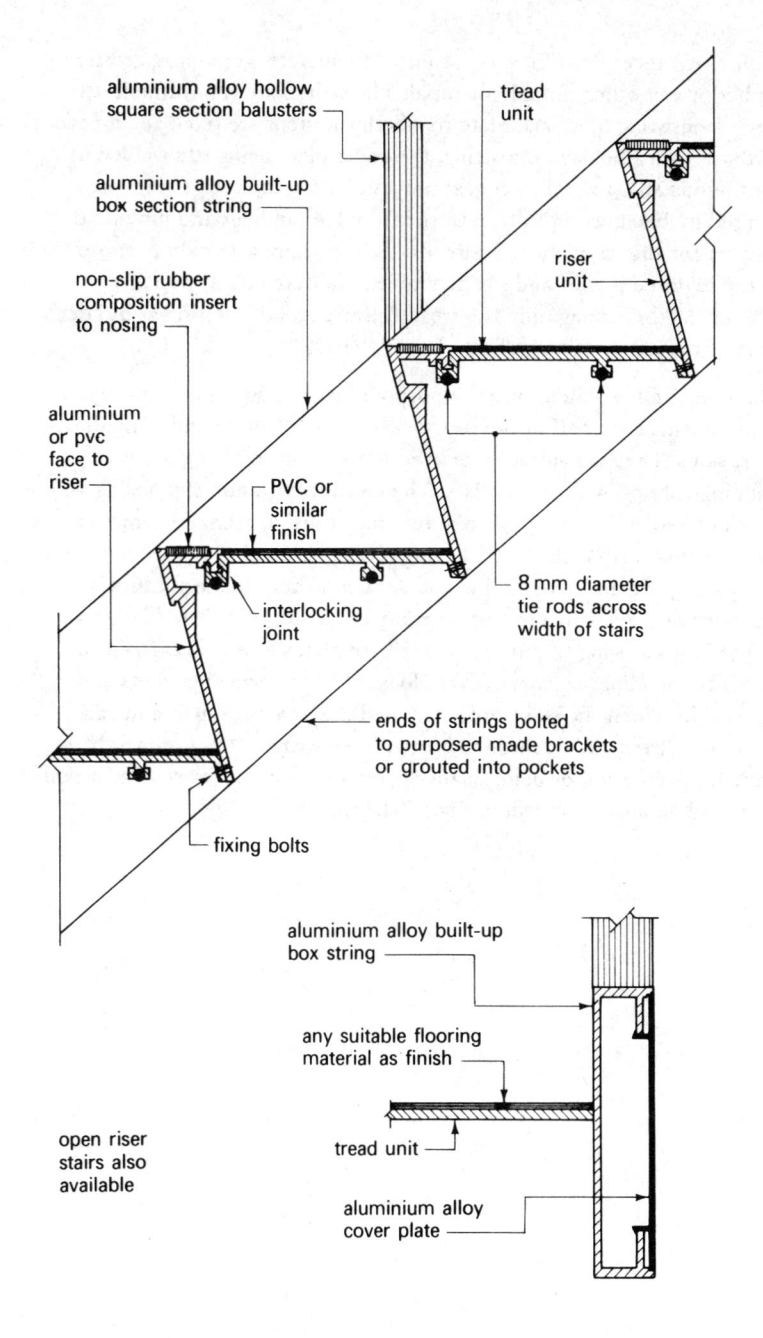

Fig VIII.15 'Gradus' aluminium alloy stairs

component recessed to receive a filling of concrete, granolithic, terrazzo, timber or any other suitable material. The strings are very often in two pieces consisting of a back plate to which the steps are fixed and a cover plate to form a box section string, the cover plate being site welded using a continuous MIG (metal inert gas) process. The completed strings are secured by brackets or built in to the floors or landings and provide the support for the balustrade. Stairs of this nature are generally purpose made to the required layout and site assembled and fixed by a specialist sub-contractor leaving only the tread finishes and decoration as builder's work — typical details are shown in Fig. VIII.14.

Aluminium alloy stairs: usually purpose made to suit individual layout requirements with half or quarter space landings from aluminium alloy extrusions. They are suitable for accommodation stairs in public buildings, shops, offices and flats. The treads have a non-slip nosing with a general tread covering of any suitable floor finish material. Format can be open or closed riser, the latter having greater strength. The two part box strings support the balustrading and are connected to one another by small diameter tie rods which in turn support the tread units. The flights are secured by screwing to purpose made base plates or brackets fixed to floors and landings or alternatively located in preformed pockets and grouted in. When the stairs are assembled they are very light and can usually be lifted and positioned by two men without the need for lifting gear. No decoration or maintenance is required except for routine cleaning — typical details are shown in Fig. VIII.15.

Bibliography

Relevant BS — British Standards Institution.
Relevant BSCP — British Standards Institution,
Building Regulations 1985 — HMSO.
Relevant BRE Digests — HMSO.
Relevant advisory leaflets — DOE.
DOE Construction Issues 1—13 — DOE.
R. Barry. *The Construction of Buildings*. Crosby Lockwood & Sons Ltd.
Mitchells Building Construction Series. B. T. Batsford Ltd.
W. B. McKay. *Building Construction*, Vols. 1 to 4. Longman
  Specification. The Architectural Press.
R. Llewelyn Davies and D. J. Petty. *Building Elements*. The Architectural Press.
Cecil C. Handisyde. *Building Materials*. The Architectural Press.
W. Fisher Cassie and J. H. Napper, *Structure in Buildings*. The Architectural Press.
Handbook on Structural Steelwork — The British Constructional Steel Work Association Ltd and The Constructional Steel Research and Development Organisation.
L. V. Leech. *Structural Steelwork for Students*. Butterworths.
B. Boughton, *Reinforced Concrete Detailers Manual,* Crosby Lockwood & Sons Ltd.
T. Whitaker. *The Design of Piled Foundations*. Pergamon Press.
A. S. West. *Piling Practice*. Butterworths.
Fire Protection Series by E. L. Wolley. *Building Trades Journal.* Northwood Publications.

Drained Joints in Precast Concrete Cladding. The National Buildings Agency.
Relevant A J Handbooks. The Architectural Press.
Data Sheets — British Precast Concrete Federation.
Relevant manufacturers' catalogues contained in the Barbour Index and Building Products Index Libraries.

Index

A

Accommodation stairs, 215, 231–2, 236–8
Aggregate notional area, 115
Aggregate transfer, 212–13
Aluminium alloy sheets, 179
Aluminium alloy stairs, 237–8
Asbestos cement cladding, 168, 179
   roof, 168
   wall, 179
Asbestos insulating boards, 109–10
Auger boring, 20, 56, 58–9
Automatic fire doors, 137–8, 174
Automatic roof ventilators, 175–6

B

Baffle strips, 153–8
Basement storeys, 98–9
Basement timbering, 7, 9
Basements, 6–8
Bending moment diagrams, 67
Bentonite, 11–13
Boundaries, 102
BSP cased piles, 48–9
BSP prestcore pile, 44, 60–1
Building Regulations, 98–116, 118–19
Bush hammering, 212
Butyl mastics, 161

C

Cantilever stairs, 218, 222–3
Carbon dioxide, 118
Carbon monoxide, 118
Cased piles, 48–9
Cast-in place piles, 48–52
Cast-on finishes, 213
Cheshire auger, 56, 58
Cladding panels, 141–6
   concrete, 142–6, 158
Claddings, 141–62
Climbing formwork, 190, 193
Coated steel sheets, 179
Cofferdams, 8, 10
Coloured cement, 208
Column underpinning, 39-40
Compartment floors, 98–9
Compartment walls, 98–9, 101
Compartments, 98
Composite piles, 44, 47
Concrete floors, 112
Concrete portal frames, 69–75
Concrete surface finishes, 208–14
   colour, 208
   texture, 209
Construction Regulations, 1–2, 15, 26
Cranked slab stairs, 218, 221, 225, 227
Cutting shield, 20–2

241

D

Daylight factor, 164
Deep trench excavations, 2–3
Deep trench timbering, 3
Demolition, 24–9
  bursters, 28
  deliberate collapse, 27
  demolition ball, 27
  explosives, 28
  hand, 26
  pusher arm, 26–7
  statutory notices, 25
  survey, 24
  thermal reaction, 28
  thermic lance, 28–9
  wire rope pulling, 28
Diaphragm walls, 8, 11–13
Diesel hammer, 52, 54–5
Displacement piles, 43–52
Doors, 134, 136–8
  automatic fire, 137–8
  fire check, 134, 137
  fire resisting, 136–7
  smoke stop, 127, 137
Double acting hammer, 52
Downdrag, 42–3
Drained joints, 143–4, 153–9
Driven *in situ* piles, 48, 50–1
Drop hammer, 52

E

Efflorescence, 209
Elements of structure, 100
Enclosing rectangle, 114
End bearing piles, 42
Escape routes, 119, 130
Escape stairs, 134–5, 137, 139, 215
Excavations, 2–20
  basements, 6–8
  cofferdams, 8, 10
  deep trench, 2–3
  diaphragm walls, 8, 11–13
  shafts, 15–18
  tunnels, 16, 18–20
Exposed aggregates, 213
External escape stairs, 139–40

F

Factories Act 1961, 174
Factory buildings, 163–87
  roofs, 163–76
  walls, 177–82
  wind pressures, 183–7

Filled joints, 152–3, 155
Final pinning, 37
Fire, 87–140
  Building Regulations
    Part E, 98–116
    Part EE, 118–19
  doors, 134, 136–8
  means of escape, 117–40
  problems, 87–9
  structural fire protection, 90–116
Fire load, 90–1
Fire Precautions Act 1971, 120–1
Fire resistance, 91–2, 105–13
Fire tests, 92–7
Fire stopping, 101
Flat jacks, 37
Flats – means of escape, 121, 125–8
Flight auger, 56, 58
Formwork, 188–214
  beam, 203
  column, 203
  patent, 199–201
  slab, 205
  table, 201, 206–7
  wall, 188–98
Foundations, 30–41, 70–1, 76–7, 80
Frameworks, 65–86
Friction bearing piles, 42

G

Gas expansion burster, 28
Glass fibre reinforced plastics, 146
Glued laminated frames, 81–3
Grinding concrete, 213
Ground beam, 63–4
Gust speeds, 183

H

Hand demolition, 26
Hardrive precast pile, 45
Heat loss, 172
Hinges, 66–8, 74, 79
Horizontal casting, 213
Horizontal joints, 154, 157–8
Hotels – means of escape, 121–4
Hydraulic burster, 28

I

Inclined slab stair, 216, 218–19

Infill panels, 147–51
   arrangements, 147–8
   metal, 150
   timber, 149
Intumescent strip, 136

J
Jacking rods, 195–6
Jack pile, 34–5
Jointing, 152–9
   drained, 153–9
   filled, 152–3, 155

K
Kicker, 189, 191, 193, 195, 204
'Kwikform' table forms, 206–7

L
Lattice girder, 165–6, 169–71
Lavoisier, Antoine, 87
Lightweight decking, 171
Lightweight wall claddings, 178–82
Loop tie, 190, 193
Louvred fire ventilator, 175

M
Maisonettes – means of escape, 121, 126–9
Mastics, 160–1
Means of escape, 117–40
   Building Regulations, 118
   flats and maisonettes, 121, 126–9
   hotels, 121–4
   offices and shops, 126, 130–4
   planning, 119
   staircases, 134–5
Metal cladding, 180–1
Metal decking, 171, 181
Metal infill panel, 150
Metal stairs, 231–8
   aluminium alloy, 237–8
   escape stairs, 139, 232–3
   pressed steel, 232, 236, 238
   spiral stairs, 232, 234
   string beam, 232, 235
Miga pile, 34–5
Monitor roofs, 165, 169–71

N
Needle and pile underpinning, 34, 36
Northlight roofs, 165–8
Notional boundaries, 102

O
Offices – means of escape, 126, 130–2, 134
Oil-bound mastics, 161
Open riser stairs, 225, 228
Over purlin insulation, 172–3

P
Patent formwork, 199–207
Patent glazing, 165–9, 170–1
Percussion bored piles, 54, 56–7
Permanent formwork, 197–8
Perimeter trench, 6–7
Pile caps, 62–4
Pile framing, 5
Piled foundations, 41–64
   caps, 62–4
   classification, 41–2
   contracts, 64
   displacement, 43–52
   downdrag, 42–3
   driving, 52–5
   end bearing, 42
   friction, 42
   replacement, 54, 56–61
   testing, 60–2
Piling helmets, 53
Piling rigs, 45–7, 49, 53
Pinning, 37
Pipe jacking, 20–3
   pipes, 23
Place of safety, 130
Plywood faced frames, 83–4
Polysulphide sealants, 161
Polyvinyl chloride sheets, 179
Portal frames, 65–86
   BMD, 67
   concrete, 69–75
   steel, 76–80
   theory, 65–8
   timber, 81–6
Precast concrete piles, 43
Precast concrete stairs, 222, 225–30
   cranked slab, 225, 227
   open riser, 225, 228
   spiral, 225, 229–30
   straight flight, 222, 226

243

Preformed concrete piles, 43–6
Pressed steel stairs, 232, 236, 238
Pressurised stairways, 137
'Prestcore' pile, 44, 60–1
Pretest method of underpinning, 32
Protected shaft, 99–100
Protected zone, 130
Purpose groups, 98, 104
Pusher arm demolition, 26–7
'Pynford' stooling, 34, 37–8

Q

Queen tie, 166

R

Raking struts, 6, 9
Rankine's formula, 4
Reinforced concrete stairs, 216–24
   cantilever, 218, 222–3
   cranked slab, 218, 221
   inclined slab, 216, 218–19
   spiral, 222–4
   straight flight, 216–17
   string beam, 218, 220
Release agents, 209
Relevant boundaries, 102, 113–16
Replacement piles, 54, 56–61
Rigid portal frame, 66–8, 71–3, 77–8
Rooflights, 164
Roofs, 163–76
   daylight, 163–4, 167
   fire ventilation, 174–6
   monitor, 165, 169–71
   northlight, 165–8
   thermal insulation, 169, 172–4
Rotary bored piles, 56, 58–60

S

Sandwich construction, 173
Sealants, 160–2
Shaft timbering, 17–18
Shafts, 15–18
Shear leg rig, 56–7, 61
Shops — means of escape, 126, 130–2, 134
Shot blasting, 212
Silicone rubber sealant, 161
Single acting hammer, 52–3
Site works, 1–29
Sliding formwork, 194–6, 198
Smoke, 117
Smoke logging, 174, 176
Spiral stairs, 222, 224–5, 229–30, 232, 234
Splice joints, 73, 79

Sprayed asbestos, 108
Stairs, 215–38
   in situ concrete, 215–24
   metal, 231–8
   precast concrete, 222, 225–30
Staves, 194–96
Steel portal frames, 76–80
Steel preformed piles, 44
Steel stairs, 232–8
   escape, 139, 232–3
   pressed metal, 232, 236
   spiral, 232, 234
   string beam, 232, 235
Storey height cladding panels, 142–3, 145
Straight flight stair, 216–17, 222, 226
String beam stair, 218, 220, 232, 235
System formwork, 199–207

T

Table formwork, 201, 206–7
Thermal Insulation (Industrial Buildings), 172–4
Thermal reaction demolition, 28
Thermic lance, 28–9
Three pin portal frame, 66–8, 74, 78–9
Thrust boring, 20
Thrust frame, 21–2
Timber and plywood gusset frames, 83, 85–6
Timber floor, 111
Timber infill panel, 149
Timber piles, 43
Timber portal frames, 81–6
Timbering, 2–7, 9, 17–19
Tremie pipe, 11–12
Triangle of fire, 88
Tripod rig, 56–7, 61
Tucking framing, 5
Tunnel timbering, 18–19
Tunnelling, 14–23
Tunnels, 16, 18–20
Two pin portal frame, 66–7, 74, 79

U

Underpinning, 30–40
   beams, 37
   columns, 39–40
   final pinning, 37
   jack pile, 34–5
   needle and piles, 34, 36
   pretest method, 32

'Pynford' stooling, 34, 37–8
  survey, 30–1
  wall, 31–3
Under purlin insulation, 172–3
Underreamed pile, 56–9
Underreaming tool, 59
Undersill cladding panel, 142, 144
Unprotected areas, 103, 114–15
Unprotected zone, 130

V

Vertical joints, 153–9
Vibro pile, 51
Vortex, 185

W

Wall formwork, 188–98
  climbing, 190–3
  permanent, 197–8
  sliding, 194–6
  traditional, 189–92
Wall functions, 177
Wall underpinning, 31–3
Water bar, 155–8
Water jetting, 54
Wedge bricks, 37
Wind bracing, 186–7
Wind pressures, 183–7
Wire rope pulling demolition, 28
Working temperatures, 172